Busting the Myth of the Communication Metaphor

SUNY series, Studies in Technical Communication
———
Miles A. Kimball, Derek G. Ross, and Hilary A. Sarat-St. Peter, editors

Busting the Myth of the Communication Metaphor

How Technical Writing Conventions Perpetuate Injustice

SARAH READ

Published by State University of New York Press, Albany

© 2025 State University of New York

All rights reserved

Printed in the United States of America

No part of this book may be used or reproduced in any manner whatsoever without written permission. No part of this book may be stored in a retrieval system or transmitted in any form or by any means including electronic, electrostatic, magnetic tape, mechanical, photocopying, recording, or otherwise without the prior permission in writing of the publisher.

Links to third-party websites are provided as a convenience and for informational purposes only. They do not constitute an endorsement or an approval of any of the products, services, or opinions of the organization, companies, or individuals. SUNY Press bears no responsibility for the accuracy, legality, or content of a URL, the external website, or for that of subsequent websites.

EU GPSR Authorised Representative:
Logos Europe, 9 rue Nicolas Poussin, 17000, La Rochelle, France
contact@logoseurope.eu

For information, contact State University of New York Press, Albany, NY
www.sunypress.edu

Library of Congress Cataloging-in-Publication Data

Name: Read, Sarah, author.
Title: Busting the myth of the communication metaphor : how technical writing conventions perpetuate injustice / Sarah Read.
Description: Albany : State University of New York Press, [2025] | Series: SUNY series, Studies in Technical Communication | Includes bibliographical references and index.
Identifiers: ISBN 9798855802894 (hardcover : alk. paper) | ISBN 9798855802870 (ebook) | ISBN 9798855802887 (pbk. : alk. paper)
Further information is available at the Library of Congress.

Contents

List of Illustrations		vii
Acknowledgments		ix
Preface		xi
Chapter 1	Harm Is Being Done	1
Chapter 2	Making the Communication Metaphor Visible	29
Chapter 3	What Does the Communication Metaphor Mean?	57
Chapter 4	Where Does the Communication Metaphor Come From?	89
Chapter 5	How Is the Communication Metaphor Perpetuated and Maintained?	123
Chapter 6	Experiments in Imagining a Post–Communication Metaphor World	165
Notes		195
References		213
Index		221

Illustrations

Figure I.1	The Met-o-Matic Machine	xii
Figure 2.1	Communication metaphorest fire complex	31
Figure 2.2	Cybernetic loop diagram	35
Figure 2.3	Communication black box feedback loop	36
Figure 3.1	The container metaphor	65
Figure 3.2	The conduit metaphor	66
Figure 3.3	Containers move packages across distances	66
Figure 3.4	Alternative model for communication to the conduit metaphor.	70
Figure 3.5	Line as both track and map	76
Figure 3.6	The windowpane theory of language	77
Figure 3.7	The transmission model of communication	83
Figure 3.8	The conduit metaphor	84
Figure 4.1	Detail of cork from scheme XI in Hooke's *Micrographia*	94
Figure 4.2	The transmission model of communication	120
Figure 5.1	Starbucks Coffee Company terms of use	150
Figure 6.1	Recreated Table 1 with data redacted	171
Figure 6.2	Doughnut	179

Figure 6.3	Jelly doughnut in water	179
Figure 6.4	Jelly doughnut in water filled with medicine	180
Figure 6.5	Jelly doughnut in water filled with medicine with scale	181
Figure 6.6	Liposome	182

Acknowledgments

Since this book is the product of my own research, experiences, thinking, and reflection on technical writing, it has been profoundly shaped by the people in my life who share this world with me. This world of technical writing is broad and reaches across boundaries, both real and imagined, between academia and industry, theory and real life, and professional and personal relationships.

First and foremost, this book was written for the members of the Technical Literature Saloon, a Seattle area writing group of current and former aerospace engineers, computer programmers, writers, and professionals who have welcomed my presence as "the creative writer" for twenty-four years. In particular, I am grateful to Brian Tillotson, a holder of over one hundred aerospace patents, for his insightful feedback on early drafts of this book.

I am also grateful to Doug Downs, a longtime colleague and friend, who served as a developmental editor for this book project. Who knew that you could use start-up research funds to hire a developmental editor as an accountability partner, sounding board, critical reader, and cheering section for a long book project? For anyone dreaming of writing a book, I highly recommend it.

I am thankful for current and former students, especially Jesse Johnson, who threw down the gauntlet on technical writing in the best way possible, and Lucas Schaumberg, whose informed conversation about the philosophy of language and science helped shape chapter 4.

Many colleagues in the Portland State University English Department and beyond have listened to me talk, read drafts, and offered feedback, including Jacob Tootalian, Michael Michaud, and Shawna Shapiro. My longtime collaborator Jordan Frith unknowingly contributed to my thinking as we worked collaboratively to articulate an alternative metaphor for writing: infrastructure. Thanks also to Lynn Santelmann and other colleagues in Applied Linguistics at PSU.

I'd like to thank my colleagues in science at PSU, including especially Jay Nadeau and her lab, for getting me out of my head and onto the front lines of science.

Thanks to Rori Anderson in the English Department for help formatting the manuscript, and to Oregon graphic designer and illustrator Duke Stebbins for bringing difficult abstract concepts to life with some bonus humor and whimsy.

Thank you to the anonymous reviewers engaged by SUNY Press whose comments both encouraged me to strengthen aspects of my argument while also shoring up my conviction in its value for a broad audience.

Finally, thanks to my loving and supportive family, Richard and Jane, who are also scientific and technical writers, sometimes reluctantly, and who, thankfully for me, put up with my feedback on their work with grace, most of the time.

Preface

If you are a scientist, engineer, or technical professional, or a student on your way to becoming one, then you likely agree that the following statements are true, or you have heard them quoted by a senior colleague, boss, manager, or professor and you are on your way to accepting them as part of your technical and professional identity:

1. Writing and communication are important to my work.
2. Clarity is the most important element of successful written communication.
3. Successful written communication is:
   - concise, accurate, elegant, informative, and engaging
   - accurate, relevant, easy to understand, and accessible
   - short and to the point, with the facts only, no opinions

All of these statements are direct quotes from a collection of comments about writing and communication by scientists and technical professionals whose voices echo throughout this book. These statements could have also just as easily been lifted from an introductory technical writing textbook or quoted from someone you know. In fact, it would be hard to find a scientific or technical professional or researcher, or even a person on the street, who could find strong reasons to disagree with these statements.

Collectively, these statements represent an orthodoxy that I have come to call the Communication Metaphor, which is the subject of this book. As a whole, I argue, these tacitly held beliefs and practices about what makes for successful technical writing, the Communication Metaphor, stand in as symbolic for a messier, more reality-based understanding of how writing and communication really works.

An orthodoxy is a set of widely held beliefs and practices that are accepted to be both true and useful, because they normally are true and useful, for many people most of the time. However, this same set of received wisdom is also neither true *nor* useful for some people *all* of the time; in fact, for some people the Communication Metaphor has actively caused themselves, or their communities, harm.

Over the last twenty years that I have been teaching technical writing, I have become increasingly aware of a gap between how we talk about how technical communication works and the success of its outcomes for many people, especially women, people of color, and other underrepresented groups in scientific and technical fields. This growing realization has made me thoughtfully uncomfortable with the reality that widely accepted beliefs and practices that work for many people much of the time also don't work for some people all of the time. In other words, exclusion is built into the most broadly held truisms about what makes for successful technical writing. As an academic, questioning broadly accepted and practiced beliefs and working to understand the sources and motivations behind them is

Figure I.1. The Met-o-Matic Machine turns messy reality into the comfortable Communication Metaphor.

simply what I do; however, as a practicing writer, teacher, and citizen I have become increasingly tired of witnessing how such seemingly benign, apparently beneficent, widely taught conventional knowledge does so much harm.

One of the most poignant sources of my own discomfort is some wisdom I gained several years ago from an indigenous master's student in the professional and technical writing program at Portland State University. Over the course of his two years in the program, Jesse researched an extensive bibliography on rhetorical sovereignty, which is "the inherent right and ability of peoples to determine their own communicative needs and desires in this pursuit to decide for themselves the goals, modes, styles, and languages of public discourse."[1] At his oral exam before graduation, during which we discussed his excellent long essay on a recent example of how the Tohono O'odham Nation in Arizona used the legal framework of two gaming regulatory acts to reclaim their right to develop a casino on tribal lands, Jesse made clear to me and his graduation oral exam committee that our program falls entirely within and is wholly complicit in the maintenance and propagation of Western Communication practices (his choice to capitalize). His case study of the Tohono O'odham Nation is but one example of how these communication practices have historically caused harm to Native Americans, and continue to do so in the present. Without yet going too far into the weeds of the standardization of linguistic systems and practices and how standardization consolidates power for dominant groups, it's fair to say that at a high level what Jesse meant by "Western Communication practices" is equivalent to the statements that open this preface.

Having gotten to know Jesse's work during his time in the program, his comments during his oral exam came as no surprise to me intellectually or as a scholar, but they made me personally uncomfortable in a way that has forced my continued reflection on his message, including writing this book. What Jesse helped me to understand was that despite my mentorship over his two years in the program and my encouragement of his work regarding rhetorical sovereignty, I don't get to own any of it. Certainly, Jesse owned and always owned his intellectual work that took the form of the long essay, but he also owned 100 percent the powers of agency and resistance that are implicit to the practice of rhetorical sovereignty. For Jesse, the academic exploration and development of this rhetorical strategy was a conceptual framework for practically encountering and finding agency in technical communication, a field and profession that has frequently been critiqued for how white it is.[2] As I am the director of the program, this meant that rhetorical sovereignty was also a tool to regain agency from me, who, by the nature of my position and my responsibilities to the English Department and the university, am complicit in maintaining and reproducing the current oppressive system of linguistic injustice that his tribe has suffered under.

This is to say that there is no escape clause for me here—not in mentoring a Native American student nor in learning or writing about rhetorical sovereignty. Both of these acts are useful and well-meaning acts of self-education, but without further action they leave the overall system in place. This book, then, which makes visible and critiques the Communication Metaphor with the ultimate aim to replace it, is, in part, the beginning and middle stages of my own processing and contemplation of what it means to navigate this very personal discomfort. I am also aware that I am privileged to not have to do this on my own given that my position as a white member of the field of technical communication is neither unique nor rare. In this book I invite readers—both professionals in industry and students and professors in academia—to join me, patiently, in mapping and crossing the unforgiving and rugged landscape of the Communication Metaphor and its sources and consequences.

I do want to share one other line of my thinking that has recently started to unsettle me. Chapter 4 of this book relies heavily on an excellent scholarly book[3] that traces how scientific knowledge became a dominant way of thinking in society and how technical writing developed as the primary medium—or currency—for exchanging and exploiting its value. The argument springs from the assumption that scientific and technical knowledge has historically grown to hold, and currently enjoys, a powerful, privileged place in society. As I have written this book and continued to reflect personally, I have grappled with the uncomfortable feeling that while this could possibly have been true when Bernadette Longo's book was published in 2000, just twenty-five years later I'm not sure that it still is, or possibly if it ever has been at all. After all, from a citizen's perspective, the logical end of a society dominated by scientific thinking is a more technocratic one that would have initiated scientifically informed climate change policy that would have stopped or at least slowed climate change decades ago or one that would have created the conditions to slow or even eradicate the COVID-19 pandemic via the rapid and universal uptake of vaccines. Given that neither of these happened in our very recent memory demonstrates that in these early decades of the twenty-first century, science is, in fact, likely *not* any longer the dominant paradigm for society, if it ever was. And while my purpose here is not to reignite the detailed political and policy debates of climate change and COVID-19 vaccines, both of which are issues that have deep cultural, social, and scientific meaning for society, this general line of thinking does raise for me a rather uncomfortable question as a writing scholar: Are scientists and other technical professionals already more generally writing from the margins, and if so, what does this mean about the conventions for technical writing and the Communication Metaphor? In other words, while the conventions for scientific and technical writing have historically developed to exclude those whose home language is not Jesse's Western Communication, scientific and technical ways of thinking, arguing, and writing may also and already be marginalized at this point in history. To be clear, this book is not primarily about unpacking this line of mostly undeveloped thinking; however, my sense that something fundamental has changed for science and technology's positioning in society is a powerful undertow to the argument in this book, which both implicitly, and at times explicitly, adopts Longo's overall thesis.

That the primary work of this book is definition means that the project of this book is primarily hermeneutic. Hermeneutic projects reveal and shine light on what is otherwise difficult to see. In other words, the argument in this book is a work of interpretation rather than a report on empirical data. What I offer is what scholars who primarily study texts would call a "reading" of the evidence that I have collected about the Communication Metaphor through the lenses of scholarship in technical communication, writing studies, rhetoric, and linguistics, and my own research and experiences. Because I draw on a transdisciplinary set of scholarship and other sources, the interpretation I argue for here is synthetic, or a combination of ideas to form a new holistic understanding. This new holistic understanding, what I call the Communication Metaphor, likely does not fit cleanly into a single academic discipline. However, the aim of this book is not to make a narrow argument that builds on a single thread of scholarship. Rather, the aim of this book is to make a broad argument that sketches a landscape within which many strands of scholarship, as well as personal experiences, can be located or contextualized. This is especially true for the scholarly endeavors of Black technical and professional communication and other scholarship related to linguistic justice. It is important to be clear that the argument I make here is not itself necessarily within these strands of scholarship. That is, I am not primarily a scholar of Black TPC or linguistic justice, although the urgency that drives my inquiry into the Communication Metaphor is largely motivated by recent Black TPC voices.[4] In addition, our collective awareness as a culture that systemic injustice must be dismantled requires that everyone be willing to examine the assumptions and practices that structure our daily lives. In general, I subscribe to this statement by French language philosopher and historian Michel Foucault: "The work of an intellectual is not to mold the political will of others; it is, through the analyses that he [sic] does in his [sic] own field, to re-examine evidence and assumptions, to shake up habitual ways of working and thinking, to dissipate conventional familiarities, to re-evaluate rules and institutions and . . . to participate in the formation of a political will (where he [sic] has a role as a citizen to play)."[5]

Finally, I would like to share with readers how I imagine this book can be read by the various audiences that I have written it for. First and foremost, I have aimed to write a book for technical and scientific professionals and for their students and mentees. I have not assumed that this readership already has a foundation in some of the standard terms and concepts of writing studies, technical communication, and the other fields the argument in this book draws together. For this reason, there may be more explanation of standard concepts and terms than an audience of my colleagues might expect, or desire. Alternatively, there will be moments when a casual reader may want to skim past a more academic section that is offering due diligence to scholarly activity on a topic. At the same time, I do believe that the unique, synthetic construct of the Communication Metaphor has value for my colleagues in writing studies and allied fields. Finally, I have written this book for instructors of introductory professional and technical writing courses and their students with the aim that one or several chapters of the book could stand alone as readings that support a class discussion, for example on what the word *clarity* means or where technical style standards come from and who they serve. Different moments in the book will resonate more or less depending on the audience, or audiences, that readers claim. In addition, I hope that the illustrations, done by a comic illustrator with the skill to communicate tough abstract concepts via simple line drawings and occasional whimsy, will broaden the appeal, and the accessibility, of the Communication Metaphor. Overall, for all readers, my aim is for the book to raise new questions about issues that have been treated as settled for way too long.

Chapter 1

Harm Is Being Done

Let me begin with a fact that is unlikely to surprise any reader: When I collected comments from scientists, engineers, and technical professionals about their attitudes about and practices for writing and communication, 100 percent responded "yes" to the question "Are writing and communication important to your work?"

Certainly, the unanimity in the responses is not unexpected given the topic at hand—the importance of writing and communication in technical careers. For most practical purposes in academia and industry, this agreement is broadly accepted as a good thing by faculty, researchers, and students, including myself. The value of writing and communication to scientific and technical careers has been reflected over the last couple of decades in the development of program learning outcomes that emphasize communication as an essential professional skill,[1] the proliferation of engineering and science writing textbooks and manuals, and the expansion of requirements and support for technical and professional writing courses and programs at universities.

Thinking a step further, if we return to the statement "Are writing and communication important to your work?" and focus on just the subject of the sentence, we might expect that questions more specifically about "writing and communication" would elicit more variable, possibly contrary, answers. This would be a reasonable expectation given that most people find "writing and communication" very challenging to do successfully, especially in technical and professional contexts. However, when scientists and engineers were asked to answer the question "What are the characteristics of successful written communication in your workplace or profession?" the answers to these questions, again, could be easily anticipated by readers of this book:

Successful written communication is:

- concise, accurate, elegant, informative, and engaging
- accurate, relevant, easy to understand, and accessible
- short and to the point, with the facts only, no opinions
- clear and concise wording to explain complex concepts
- clarity and brevity

In short, none of these answers comes as a surprise—which is exactly the point. What interests me about the conventionality of the responses to a question about what makes for successful technical writing is simply how uniform they actually are. Over the twenty years that I have been teaching and researching technical writing, I have become increasingly awed by the predictability of answers to questions about what makes technical writing and professional writing successful. To put this another way, I have become fascinated about why answers to this question take a form closer to reciting a catechism or primer rather than the form of unique stories about experiences, feelings, or philosophies about writing.

When I ask students on the first day of an introductory writing course, or if technical writing comes up during casual conversation at a social event inside or outside of academia, or when I am chatting with neighbors, or pretty much anywhere else, I will hear some version of the statements above put forward as knowledge about what makes for successful technical writing, or why it is important. So many years of hearing the same answers from across a very diverse group of people in terms of age, experience, professional background, and other factors has got me thinking about where this widespread agreement comes from. The nearly total agreement among North American English speakers (at a minimum) about what makes for successful technical writing has finally got me asking where these values for technical writing come from, why are they so culturally prevalent, how people learn them, and, importantly during this era of renewed energy and urgency around issues of equity and inclusion, who do they exclude, or harm?

A lack of controversy about an issue in any community can be an indicator of how highly conventional beliefs and practices are no longer being evaluated for their potential to exclude, or to do harm. In the case of technical and professional writing, commonly held beliefs and practices have not yet been systematically held up to scrutiny outside of small academic circles because there hasn't seemed to be a reason to do so by the majority of its stakeholders. After all, current conventions of technical and scientific writing and communication seem to mostly work. Scientific journals publish a continuous stream of research that advances knowledge in medicine, basic research, and engineering fields. Technical companies and research organizations publish technical reports and operational reports that persuade their stakeholders or funders to continue their support and develop patents. Engineers develop specifications for commercial airplanes that are built and fly people around the world. Countless job seekers write cover letters and resumes and get jobs in the public, nonprofit, and private sectors of the economy. For the most part, the world goes around, and, arguably, better than it used to, given the contemporary ubiquity of courses and curriculum teaching technical writing and communication.

Another reason these statements about technical writing seem uncontroversial is that for the last seventy-five years or so during the post–World War II era, they have fundamentally shaped our view of what student engineers and scientists need to learn in school and what professional engineers and scientists need to perfect once on the job. And I want to say up front, before there is any possibility that this book might be read as a direct critique of this history, that there is absolutely nothing at all wrong with teaching students in science, technology, engineering, and math (STEM) majors how to communicate well. In educational and industry locations where investment in communication curriculum for the technical professions has been smart and sustained, it has worked overall—we have at least done more good than harm when preparing students to write technical reports and workplace correspondence and to give effective oral presentations supported by a slide deck. And I say this as someone who has given twenty years of my career over to teaching introductory technical communication to STEM students and to preparing master's level students for technical writing careers in industry. In addition, my research has documented how technical and professional writing and communication are essential to the missions and outcomes of organizations, including a midsize nonprofit advocacy organization[2] and a supercomputing center at a federally funded national laboratory.[3]

And yet, despite all of the good, harm has been done—and continues to be done all around us—by the continuing dominance of the complex of tacit assumptions and practices that I call the "Communication Metaphor." This individual and structural harm has been suffered primarily by underrepresented groups of people in STEM disciplines, such as women, people of color, and other minoritized groups, who are more likely to experience that the Communication Metaphor suppresses or elides their voices and contributions. For example, the broad agreement about what makes technical writing and communication successful also suggests that what shapes how we understand and execute it comes from a singular worldview with a relatively narrow set of goals and values. As linguists know, there is a direct link between the dominant language practices of a community and power: In general, the people with the power get to determine what are the acceptable and normalized ways of communicating.

When linguists say that power and language are connected, they are arguing for a social theory of discourse[4] that establishes a direct connection between our tacit beliefs and practices, or our unconscious ideologies, and how they motivate and shape our language use and what we believe about it. This is to say that we participate in the broad agreement that successful technical writing is "clear," "concise," and "objective" because those beliefs are motivated by an underlying system of beliefs that serve the interests of groups that are in power, and who aim to stay in power. When our language—or discourse[5]—practices become "naturalized and win widespread acceptance," they function as a source of power to maintain the status quo or "cultural hegemony." Historically, and continuing into the present, this status quo has benefited primarily middle-class, white men who have been overrepresented in North America in STEM disciplines in both academia and industry. On the flip side, people who don't already feel at home in the language practices of science and technology are excluded or find the price of acceptance to be extremely high. To make this point more personal, in the preface I shared the story of Jesse, a member of the Tohono O'odham Nation in Arizona, who held me accountable as the director of a technical and professional writing program for propagating *Western Communication* (his term), which has materially caused harm to his tribe in the form of treaties with the government that took away the sovereignty over their land and other rights. Jesse is right to point out that government and legal language is of a particular kind, in that it is granted the authority to assert the rights of one party over another, whether the parties are citizens, tribes, companies, or nations. Part of what grants "official" language efficacy is the collective agreement that language with certain stylistic characteristics, for example language that contains technical jargon (scientific or legalese) and that is written from a neutral point of view (no use of "I" or "we") is the voice of an unalienable authority, which, we hope, is democratically and justly constituted but in the case of Native Americans certainly wasn't.

It's hard to disagree that the ubiquitous use of technical and scientific jargon functions to exclude audiences who are not trained or empowered within the knowledge, or cultural, domain that the writing is about. And evidence shows that the situation is getting worse. A study of 709,577 abstracts of articles published between 1881 and 2015 from 123 scientific journals showed that the readability of science is steadily decreasing.[6] It could be argued that the increase in the number of syllables per word, the number of words in a sentence, and the number of words used exclusively by scientists has been made necessary by the cumulative growth of scientific knowledge over the last 350 years of institutionalized, Western science (chapter 4 will explore this history in detail). However, low readability of scientific reports doesn't just exclude general, nontechnical audiences but has also been shown to affect the comprehension of scientists and their ability to reproduce scientific results—a crucial component of the scientific method necessary for validating findings. It's also not obvious what to do about it: Studies have shown that the "mere" presence of jargon, even when *clearly* defined and well contextualized, can produce negative effects on processing fluency directly, and on self-perceptions and engagement indirectly.[7] More, it appears, is going on in comprehension than simply backfilling a deficit in knowledge about what technical terms mean.

People are left out of science not only because science writing is dense with un- or ill-defined technical language. Feelings-as-information theory (FIT) proposes that one's feelings of knowing while processing—whether the information is easy or difficult to process—informs the relationship of the information with the self. An expectation that familiar or owned information is easier to process leads to a disengagement from the self of information that is difficult to process—it actually matters a lot who you are when it comes to comprehending jargon heavy writing, whether or not it is "clearly written."

As chapter 4 will explore in more detail, the history and style of scientific and technical writing has evolved to serve a particular agenda, one that is heavily colonial and capitalist, and to serve the people, largely male and white, who have benefitted the most. Whether or not scientific writing can do anything else without fundamentally changing is one of the motivating questions of this book.

To put it lightly, this is heavy stuff. Many, but I sincerely hope not all, readers of this book are members of groups that hold power in technical fields, or who at least benefit from one or more privileged identities, whether middle-class, male, or white. However, if the connection between ideology and language practices is too abstract for you, we can also look to more material evidence that conventional beliefs and practices about technical writing exclude and therefore cause harm. If we take as a given that those in power get to establish and maintain the norms for writing and communication, then the continuing lack of diversity regarding gender, race, class, and ethnicity across most technical professions in both academia and industry serves as direct evidence that technical language is part of the problem.

We need only look at recent employment statistics to see that conventions for language use in technical contexts are not being set by underrepresented groups: women make up only 18 percent of full-time employed software developers and only 9 percent of electrical and mechanical engineers.[8] In addition, only 5.5 percent of employed engineers aged twenty-five to thirty-four years with bachelor's degrees are black women, compared to 9.2 percent for white women and 24.1 percent for white men, the most represented demographic.[9] Fortunately, some major employers in industry have recognized that diversity in the workplace is an ongoing problem and have declared their intentions to work to remedy it. For example, in 2022 Boeing stated as a goal of its equity and inclusion initiatives to "clos[e] representation gaps for historically underrepresented groups and increas[e] the Black representation rate of our U.S. workforce by 20%."[10] It's hard to know at this point whether these efforts will be effective or sustained long enough to motivate system change. From my perspective, the answer is no, since I see the roots of the problem not just in hiring and managerial practices but in the deeper cultural and language practices of technical professions. At this point some readers may want to point to themselves or their colleagues as evidence that scientists and engineers who do not claim privileged identities can in fact be hugely successful in STEM careers. And this is true. The argument of this book will not necessarily reflect the experience of any given individual, but it does aim to make a useful point about the sea that we all swim in. In addition, we want to make sure that we notice who continues to be missing from or underrepresented in technical fields and why by not deflecting our attention away from those folks by focusing on who has made it and how.

We can also look to examples of how language practices are widely regulated in order to maintain the status quo that clear, concise, and objective writing is equivalent to credible, authoritative knowledge. If this claim seems initially too abstract and removed from our daily experience, I suggest considering the stated purpose of one of the most frequently visited sites on the internet, Wikipedia: "Wikipedia's purpose is to benefit readers by acting as a widely accessible and free encyclopedia; a comprehensive written compendium that contains information on all branches of knowledge. The goal of a Wikipedia article is to present a neutrally written summary of existing mainstream knowledge in a fair and accurate manner with a straightforward, 'just-the-facts style.' Articles should have an encyclopedic style with a formal tone instead of essay-like, argumentative, promotional or opinionated writing."[11] At first glance, this purpose statement may seem uncontroversial—Wikipedia is after all an encyclopedia that we expect to be an accessible compendium of general knowledge. But a more careful read of this purpose statement reveals contradictions that actively, and purposefully, define what counts as knowledge and what does not make the cut: on the one hand, Wikipedia contains information on all branches of knowledge, yet on the other hand, this knowledge must be neutrally written (or "objective"). The knowledge must be mainstream and suitable to a just-the-facts style of presentation, wherein a fact must be citable to a "reliable, published source." These reliable, published sources are most often going to be products of the vast scientific and academic publishing industry in English, which also, generally, adheres to similar values for what counts as knowledge. Simply put, if your knowledge isn't published in an established journal, then it can't be established as a "fact."

While Wikipedia is generally not where academic and professional scientists and engineers go first for current and credible new knowledge in their field, it is where students and broader society goes for summaries of general knowledge, including scientific and technical topics. Given its growing ubiquity and popularity (which, in my opinion, is not a bad thing), it is fair to assume that what Wikipedia declares as counting as knowledge is not uncontroversial to the general population. Whether there is general awareness of it or not, Wikipedia's purpose statement excludes branches of knowledge that are not verifiable via publication in scientific or academic journals. Wikipedia's purpose statement is an excellent illustration of how stylistic norms for writing, such as "neutrality" and "objectivity," also police what counts as knowledge. For example, indigenous knowledges have been documented as having a hard time finding purchase on Wikipedia[12] because Western norms of neutrality and objectivity do not determine the credibility or value of a finding or a narrative.

To continue to build the case that there is a problem with how we conventionally believe technical writing and communication to work, let's return to the fact that 100 percent of respondents answered "yes" to the question "Are writing and communication important to your work?" While there is broad agreement about the positive value of "technical writing and communication" for technical careers, what the finding means depends on what we understand "writing and communication" to actually be. Assuming for the time being that writing is an aspect of communication, what this statement means depends on what the large word *communication* means. This is a huge question. In sum, I argue that the answer is the Communication Metaphor, the subject of this book. However, the word *communication* initially deserves a red flag because, quite independent of the context of this particular discussion, it is a nominalization. Any style guide or primer for technical writing will tell you that a nominalization is a noun form (communication) of a verb (to communicate). Conventional style wisdom would advise that in the interest of readability and transparency, nominalizations should be avoided in technical writing. Yet, *communication* is a very common word in technical and professional writing textbooks, in curriculum, and in the talk about language use in technical fields. In addition, nominalizations in general are common in scientific and other technical writing because they are useful, as will be discussed at length in chapter 4. So, what should we be concerned about?

Nominalizations are useful because they are really handy for stuffing complex, currently or formerly controversial, abstract notions into a single word for the purpose of conserving space and reader cognitive power and creating an audience that, hopefully, understands the gubbins of what has been stuffed into the single, nominalized word. In short, the power of a nominalization is that it can substitute as a single word for what would otherwise be a book-long technical or theoretical treatise detailing the various competing views on a topic.

When it comes to the word *communication*, there are, not surprisingly, many such treatises full of controversies, which I will get to later in this book. But what is important now is becoming aware of the consequences of the substitution of a book-long treatise by a single word. Without knowing what is meant by *communication*, we simply can't know what is "important" about it for scientists and engineers. In the absence of all of that omitted information, writers, readers, and speakers will apply their own, likely unconsciously held, assumptions and values for what the word means—which can have both positive and harmful consequences. In a very general sense, the substitution of the single word *communication* for book-long treatises and all of the baggage that writers, readers, and speakers apply to it is what I am talking about when I talk about the Communication Metaphor.

This book is about the Communication Metaphor, and this chapter aims to establish the problem that warrants spending the time to flesh out the fine detail. For now, let me continue to establish the Communication Metaphor as a thing that shapes all of our experiences with technical language. I'll do this with four anecdotes that begin to illustrate it in action and how it does harm. The first is a recollection of a repeated personal experience that I have had over many years of teaching and researching technical writing. The second is an excerpt from the most commonly adopted standard textbook for technical writing. The third consists of the comments of an instructor of technical writing at a tribal college in the upper Midwest. The fourth is a collection of quotes from people who work in technical environments about what makes for successful technical writing. In each case, an underlying conception of what "communication" is and what makes it successful has consequences for the situation, often harmful ones, although not necessarily explicitly.

"Better You Than Me!"

I have been a teacher and a scholar of technical writing at five different institutions of higher education as a graduate student, an instructor, and a tenured professor. For twenty years at both university and community functions I have been asked what I do professionally. This is not in any way unusual; however, over the last ten of these twenty years I have been increasingly amazed by—and come to dread—the predictability of the course of the conversation. Over these years I have had some form of the exchange sketched below while in line for the buffet table at a faculty networking or conference event or at a cocktail party within my own social network.

POLITE CONVERSATIONALIST: And what do you do?

ME: I am a professor at Portland State University.

POLITE CONVERSATIONALIST: Oh, how interesting! [often accompanied by a gender- and age-based eyebrow-raising-in-surprise microaggression as the polite conversationalist rearranges in their mind that I am not a graduate student or stay-at-home mom and that I am in fact a highly trained specialist with a job that comes with a great deal of social capital]. And what do you teach?

ME: My area of research is technical communication. I also teach courses in our technical writing master's program and I coordinate the introductory technical writing courses taken by engineers and other STEM students.

POLITE CONVERSATIONALIST: Oh, that is such important work! I have a friend who is an engineer and she complains to me all the time about how new engineers at her firm can't write—their grammar is just awful! It is so important that engineers are able to communicate as part of their jobs. I wish our schools did a better job of teaching them how to write. I don't envy you that work, however—whew!! [followed by a sympathetic look, soft chuckles, etc.].

ME: [Deciding whether I want to get into a deeper conversation about what we actually teach those engineers beyond grammar or to reframe the conversation around my scholarly activity, but landing on the fact that I'm not up to having that conversation for the hundredth time] "Yes, I do find it satisfying work. I really like working with engineering students. What do you do?"

Despite the relative banality of this exchange, which normally ends with no social tension and the polite turn to focus on my conversation partner's occupation, it is one that I always experience negatively and have, in large part, come to dread. Why is this? Apart from the sting of what I now understand as a microaggression, which is related to my female gender, small size, and (now rapidly receding) relatively youthful look, I also think it is about the suffocating, silencing, and oh-so-inevitable narrow logic of this conversation: professor=teaching=technical communication=engineers=bad grammar=bad at communication=unsuccessful engineers=needs fixing=hero-grammar-teaching technical writing instructor.

In this scenario you can see at work the 100 percent agreement that writing and communication are important to technical careers. My conversation partner even came right out and declared it herself. For this discussion, she also took it a step further, which was to equate success with applying standardized English grammar to writing as the key to success as a communicator. This is one of the most ubiquitous logics that comprises the Communication Metaphor, a conceptual, linguistic, and, ultimately, ideological construct that powerfully shapes how we talk about and understand the purpose of technical writing and how and why it is taught to students in technical disciplines. In short, it is the logic that poor technical writing, and poor skill with communication in general, can be identified by assessing it against a grammatical and stylistic standard for good writing and communication, which may or may not be explicitly identified at the time of assessment (and usually is not).

In most contexts, including the conversation I shared above, the unacknowledged standard for written English is some version of what linguists refer to as Standard American English (SAE), Academic English (AE), or Mainstream American English (MAE), which, when they are acknowledged as a standard that has been set and enforced by a dominant racial group, are now also referred to as White English Vernacular (WEV) or White Mainstream English (WME).[13] The high value placed on the existence and efficacy of a standard form of English will be developed in chapter 5 as the Standard English fairy tale, a powerful element of the Communication Metaphor. Linguists, however, don't actually believe that standard forms of languages exist, at least not in any unchanging or neutral sense. Standard languages exist primarily in the handbooks and writing textbooks that document and propagate them. For example, for many US academic, professional, and even technical writers, the standards for written English have been and continue to be defined and enforced by the legacy of Strunk and White's bestselling *Elements of Style* (first published in 1935, with subsequent editions in 1959, 1972, and 2000). Their advice for achieving the gold standard in communication (i.e., "to Be Clear") echoes through most contemporary technical writing handbooks and manuals as well. Arguably, the unwritten code behind the advice of writing handbooks such as these is to *be clear* for middle-class white people and, conversely, unclear for anyone not nearly fluent in SAE/MAE/WME. In fact, recent critiques of Strunk and White's handbook have explicitly traced their grammar and style advice, which is not considered linguistically rigorous,[14] to the early and mid-twentieth-century social anxieties of the rising middle classes and the desire to reinforce social strata via written language use.[15]

Scholarly concern with language and injustice falls under the general banners of raciolinguistics and linguistic justice scholarship. In the field of technical communication, this area of scholarship and activism as it relates to race has recently coalesced around scholars associated with Black professional and technical writing (Black TPC).[16] Black TPC has called our attention as writing scholars and teachers back to the fact that even though *since 1974* our main professional organization, the National Council of the Teachers of English (NCTE), has espoused the value that Students' Right to Their Own Language (SRTOL)[17] in the writing classroom and in academia, conformity to SAE (or WEV or WME) continues to be the standard approach in professional and technical writing courses. The SRTOL statement declares that, first, there is no standard American dialect, only the myth of one, and, second, that the claim that any one dialect is unacceptable amounts to an attempt of one social group, whether defined by gender, race, class, ethnicity, or culture, to exert its dominance over another. It's important to point out here that this statement doesn't necessarily call for a chaotic world where there are no agreed-upon language conventions at all; however it is calling for a world where a job candidate who says or writes a non-SAE grammatical construction that is called out as an error by grammar checkers, such as in Google Documents (e.g., "I should have went to college closer to home" or "I been ready for a career change for awhile"), wouldn't be judged as unprofessional or unqualified for the job.

It should come as no surprise to almost anyone who has taught or taken a college-level professional or technical writing course that these two values have generally not shaped professional and technical writing curriculum. In short, most people agree that common sense tells us that to be clear and to be credible as writers demands a common, or standard, style. And while this common sense is not all wrong, because it has worked so well for so many people for so long, it is also not right, in the sense that injustice has been, and continues to be, done because the playing field is not level when it comes to being able to speak and write in your own language in technical and professional contexts.

While some progress has been made in writing classrooms around supporting students who do not speak or write in Standard American English or White English Vernacular[18] in ways that support the language use norms of their home communities (what scholars call "discourse communities"[19]), for the most part professional and technical writing classrooms have been considered exempt from accountability to SRTOL. After all, what is wrong with teaching *all* students how to speak or write correctly in job interviews, cover letters, workplace correspondence, and technical reporting? Today, Black TPC scholars are questioning the relevance of the 1974 NCTE SRTOL statement to professional and technical writing. They argue that curriculum that supports a student's right to their own language has been withheld from textbooks and pedagogy for engineering, science, and other professions due to what has been accepted as the benign value of teaching students from backgrounds with so-called nonstandard dialects, such as African American Vernacular English (AAVE), to code-switch to SAE or WEV.[20] In other words, the belief has been that no harm is done when teaching students who do not already write or speak in SAE, WEV, or WME to write a standard form of English in order to gain access to and acceptance in the workplace and the broader American economy. In fact, this practice is widely considered well-intentioned and for the benefit of the students as they aim to achieve social and economic mobility. This belief is easy to maintain because the Communication Metaphor enjoys such broad agreement. But what might an alternative be that is more considered, critical, and conscientious than the Communication Metaphor? By the end of this book, I hope to have offered the reader a glimpse of an alternative.

When it comes to race, linguistic justice scholars[21] are now pointing out loud and clear how assuming that all students benefit from a well-intentioned education in Standard American English is an example of white supremacy's mission to perpetuate itself. So-called standard Englishes are in fact a by-product of historical white dominance given that linguists know that no standard form of English actually exists. Linguistic justice scholars are arguing that what is called the code-switching approach to accommodating students who speak and write in nonstandard dialects into the mainstream American economy is wrong. Harm, in fact, is done when students are taught to strategically alternate between their home dialects and a standard one[22] as they move between the private (home) and the public (school, work) spheres of life. This harm takes many forms, including the violence to an individual's identity when required to suppress their home language in public and workplace settings, as well as the cultural violence inherent to the forced complicity of non-SAE communities in maintaining the marginalization of their own language. Finally, and possibly most significantly in a systemic sense, language standardization limits the types of knowledge that can be acknowledged as credible within a particular forum, such as a peer-reviewed journal or Wikipedia.

To head off a common anxiety that this argument taps into, it's important to point out that the critique of the value of a standardized language and style in technical and professional spheres does not necessarily mean that *all* language conventions are bad or not useful. Rather, it's a question of what those language conventions are and who they serve. For some readers of this book, the equation of the standardization of technical language with violence may be hard to accept, at least initially. To call this harm a kind of violence to individuals or to cultural communities can be very difficult for speakers and writers of acknowledged standard forms of English to understand. Most middle-class, white Americans can't even imagine having to code-switch, or having to significantly change their language identities, when moving among home, school, and work. Thanks to how we have benefited from the privilege that our home ways of writing and speaking (e.g., White Mainstream English) are also already accepted as academic and workplace ways of writing and speaking, we have been shielded from this kind of violence against our language and our culture simply because our language and culture has been maintained as the dominant one in academia and industry. In other words, just because we might not see or feel violence related to language use in our workplaces or schools doesn't mean that it isn't happening all the time. More metaphorically, just because we don't normally (outside of more severe earthquakes) feel the movement of tectonic plates shaping continents and pushing up mountains doesn't mean that geological forces with crushing power aren't at work all of the time.

So, given this call by linguistic justice and Black TPC scholars to revisit our assumptions about the language practices that we teach in professional and technical writing classrooms, how does this make me feel about the logic that underlies the oh-so-common conversation related above that professor=teaching=technical communication=engineers=bad grammar=bad at communication=unsuccessful engineers=needs fixing=hero-standard-grammar-teaching technical writing instructor? This logic means that my job—even my moral duty—is to remedy any failures of the communicator and their communications to adhere to the grammatical standards of SAE/WME/WEV so that both can be accepted as credible and authoritative in technical professional settings. Given the calls by the Black TPC scholars discussed above, this role also makes me complicit in the maintenance of a white supremacist language system, understood as a complicity in the continuation of a system that ensures the success of one group by marginalizing and oppressing others. Given how my conversation partner imagines my professional life based on her own experience and assumptions about what "communication" means, is this the career that I signed up for all those years ago? Even as the commonly conferred hero status has been a source of job security (for which I am grateful), let me tell it to you straight and up front—hell no! Forgive me for feeling a tiny bit of umbrage at having my professional expertise and experience essentialized into that of a finger-wagging grammar maven complicit in maintaining the language-based injustices of the Communication Metaphor!

"Incorrect Writing Makes You Look Unprofessional"

The next illustration of the Communication Metaphor at work is an initially forgettable quote from the most commonly used textbook for college-level introductory technical writing courses.[23] It serves as another illustration of how the harm done by the Communication Metaphor is hidden within broadly held, uncontroversial cultural agreements: "A correct document is one that adheres to the conventions of grammar, punctuation, spelling, mechanics, and usage. Sometimes, incorrect writing can confuse readers or even make your writing inaccurate. The more typical problem, however, is that incorrect writing makes you look unprofessional. . . . If readers doubt your professionalism, they will be less likely to accept your conclusions or follow your recommendations."[24] Few outside of a narrow scholarly community versed in language scholarship or critical theory would find any fault with the above "Measure of Excellence for Technical Documents" quoted from one of the best-selling standard introductory technical writing textbooks. The cultural commonplace is clearly stated: correct language use is credible and authoritative, and therefore professional, communication. This commonplace is repeated in many grade school or high school English or writing courses, as well as college-level introductory and professional writing courses. I would challenge the reader to find anyone outside of language study or teaching who would see immediate fault with this statement. For all its appearances, it is a completely reasonable statement that reflects the professional experience of many people, including many readers of this book. Coming to a place of understanding about the fault in this statement is one of the aims of this book.

I need only recall a recent experience that I had as a member of a board of a nonprofit community organization to know how widely normalized the cultural commonplace of correct writing=authoritative and credible communication is. In the process of drafting a statement about anti-racism for our organization in the wake of the murder of George Floyd in the summer of 2020, the committee got sidelined by a polite, but heated, multiday debate over whether a key sentence in the statement could end with a preposition. While the structure of the sentence merited ending the sentence with a preposition such that it would be structured to be most similar to how people would actually say and understand the sentence, several highly educated, professional, and extremely well-meaning members of the board remained committed to what they had learned in school, that sentences must never end with a preposition and those that do weaken the credibility and authority of the statement. For the record, even the 2000 version of Strunk and White advises that prepositions at the end of a sentence are now a matter of "your ear."[25]

To make a long story short, the debate was settled by myself and another language scholar on the board who insisted that using language the way people will best understand it is more important than adhering to traditional rules of correctness. However, the length and intensity of the debate, in the context of writing a statement with a message that was also controversial and uncomfortable for many members of the board, underscored how tightly bound correctness and credibility are in our basic assumptions about effective communication in professional contexts. The equation of correctness, as defined by adhering to a standard of language use, usually some form of Standard American English or White English Vernacular or White Mainstream English, with credibility and authority, warranted by a seemingly unalienable pragmatism demanding "clear" communication, is a characteristic of the Communication Metaphor for writing and language use. This characteristic is so normalized that to challenge it produces social tension and controversy. But it is not only the social tension generated when the correctness commonplace is challenged that matters to frame this critique—much more is at stake than that.

The quote above exposes the great deal of harm that commonplace sentiments about correctness have long promulgated for marginalized language communities. To be frank, I propose that what appears to be a sensible textbook statement about smart professional practice is in effect a thinly veiled racist statement—and I am not the first to take this position.[26] This is not to say that the author of this passage *intended* to be racist, but that the author, committed uncritically to the Communication Metaphor, simply furthered broad cultural agreements without a deeper consideration of their consequences. Because of the deep commitment to the cultural commonplace "correctness" that the author can assume to be held by the reader of this textbook, the statement doesn't have to come right out and say that certain dialects, such as African American Vernacular English, or Black English, fall outside of the standards of correctness based of SAE/

WEV/WME. The statement can imply, but not say explicitly, that so-called nonstandard dialects are, therefore, untrustworthy and noncredible, if not outright unintelligent and inferior, in professional and technical contexts. This statement reflects reality for many job applicants: A study interviewing hiring managers showed that 79 percent consider incorrect grammar, typos, and slang as markers of unprofessionalism that affect hiring decisions.[27] As will be developed in chapter 5, most people cannot confidently differentiate between a real grammatical error and a stylistic error (which includes use of so-called slang). What this means is that what most people identify as error in writing is not even linguistic but stylistic. Stylistic norms are cultural inheritances rooted in histories of power, oppression, and expediency that function as gatekeepers to the economy and are used by people in power, such as the heavily biased hiring managers.

As long as White Mainstream English is the unacknowledged norm of "official" or "technical" or "professional" language practices,[28] in both classrooms and in industry, then correctness, or adherence to these language norms, will continue to be the goal of education and professionalization efforts. In fact, this textbook statement is but one manifestation of how correctness, as just one of many aspects of the Communication Metaphor, is the product of and is sustained by a normalized systemic racism rooted in our mostly deeply held assumptions and values for language use in technical and professional contexts.

In these times when the voices of Black Lives Matter have made it into the mainstream media, albeit precariously, and many organizations in academia and industry have published statements professing values in alignment with a vision for a more just, anti-racist society—so far with uncertain levels of commitment to these values over the long haul—starting a critique of the Communication Metaphor with the issue of race is not just kairotic, or timely, it is morally imperative. In fact, this is work that needs to be done to make justice-oriented initiatives in academia and industry more meaningful because diverse language practices are not normally mentioned in company statements about work in this area. For example, the BP (British Petroleum) website "Diversity, equity and inclusion at bp" stated in 2022 that in regard to ethnicity (race is not mentioned) at the company, "What matters to us is your ambition, determination, expertise and passion—bp will be a place of opportunity regardless of ethnic identity."[29] While this is a good, if limited, overall sentiment, the statement also elides a key problem for a global, diverse, company—how do employee diversity and language diversity coexist to create opportunities rather than barriers for all? Critiquing the Communication

Metaphor is morally imperative because eliminating a source of harm for marginalized communities must, first and foremost, be our collective objective. To do this we have to understand better how the Communication Metaphor works, what are the interests in sustaining it, and how has it been accepted as a normalized commonplace for language use for so long.

"I Was Teaching Kids Who Grew Up on a Reservation, and in This Book There Are Assignments That Ask Them to Pretend Like They Are the CEO of Exxon, or Something . . ."

This third illustration of the Communication Metaphor at work is more explicit about how broadly held commonplaces about what makes for successful technical and professional communication are marginalizing to many communities. The quote in the section title comes from an interview[30] I conducted with an instructor at a tribal college in the Upper Midwest. This interview was part of a national research study I conducted during 2014–15 with my colleague Michael Michaud to document who teaches professional and communication courses at two-year, four-year, trade, and tribal institutions of higher education in the United States. We wanted to know about instructors' training and background in professional and technical writing (PTW), their approach to teaching and the nature of their employment and position at their institutions. The goal of the study was to document the ways that the teaching of professional and technical writing in American postsecondary institutions is generally underresourced, given that most instructors have contingent, part-time employment, have access to few professional development opportunities, and in some cases have little experience with technical and professional writing in industry contexts.

One of the aims of the study was to reach out to instructors at institutions that are generally not on the radar of the largest professional communities of technical communication instruction, such as the Associated Teachers of Technical Writing and the Conference on College Composition and Communication, because they teach at isolated, under-resourced institutions. Since these instructors are also not likely to attend conferences or pay close attention to professional email lists, our research assistant for the national survey and interview study searched the internet to find names of instructors at tribal colleges who might teach professional or technical writing. She reached out directly to those instructors and invited them to participate in our study, which offered only a small incentive in the form of a gift card. Thanks to the effort of the research assistant's search and the generosity of an instructor at a tribal college in the Upper Midwest, who I'll call Ilsa, we were able to document a voice about standard professional and technical writing textbooks and curriculum that cut right to the heart of the harm caused by the Communication Metaphor.

Ilsa, who is white, has taught writing and communication courses at a tribal college for many years. She talked to us about how she directly encounters the cultural conflict between mainstream white American and Native American and tribal culture in the standard professional and technical writing textbook required by her department. She let me in on how Native students are both geographically and culturally isolated from mainstream business culture, and they do not grow up with the assumption that they will join it someday—in fact, they often express skepticism about its value. In response, Ilsa limits her use of the textbook to a resource of sample documents and instead structures her curriculum around opportunities for writing projects within the local community that are engaging and empowering for students. By asking her students to write for and with their own community, Ilsa is able to teach the general principles of successful communication without asking students to imagine themselves into workplace contexts or to take on language conventions that are foreign, or even hostile to, their own culture and communities. She is not asking them to abandon culturally specific language practices that are shaped by the community's worldview, history, ideology, and language conventions that may be discouraged by the mainstream textbooks that uncritically invoke an outside standard to comply with the idea that "a correct document is one that adheres to the conventions of grammar, punctuation, spelling, mechanics, and usage . . . incorrect writing makes you look unprofessional."

While asking white mainstream college students to imagine working and writing for industry within the mainstream American economy may seem practical and uncontroversial, the same is not true for student populations who have grown up geographically, culturally, ideologically, and often socioeconomically isolated from and, quite probably, actively persecuted by it. Rereading the quotation about correctness in the second illustration via the perspective of Ilsa's students at the tribal college should make everyone feel uneasy. All of a sudden, a commonplace statement linking adherence to SAE or WEV with professional success sounds, and for many feels like, a direct cultural critique and a threat—change your language and your culture or you have no place in the American workplace or the American economy. A long history of violence against Native Americans has proceeded from this very threat, whether explicitly or tacitly enforced by government agencies, teachers, and industry leaders. After all, Native American tribes lost their rights to the lands they historically held sovereignty over via a history of treaties, agreements, and compacts that were written in English, the language of the US government—in short, professional writing in English has an approximately two-hundred-hundred-year history of functioning as a direct tool of oppression against the tribes sovereignty. Within this historical context, their suspicion of professional and technical documents written in English legalese or Standard American English is actually a survival strategy.

There are multiple possible responses to the direct threat against native langauge practices that are so mainstreamed in PTW textbooks and curriculums. Ilsa's strategy was to focus students' writing projects on local needs so that students could learn how to communicate in ways that built up their home communities. This response is very practical and highly appropriate for an introductory PTW course with the aim to teach general rhetorical skills and principles. My aim here is to illustrate the problem the Communication Metaphor creates for students not raised in the environment of Standard American English.

Everyone Says That *Clarity* Is All Important—What Is Clarity?

As a teacher of professional and technical writing I have often asked students on the first day of class what makes for successful communication. The most common answer? A statement including the word *clarity*. The statements I hear so predictably in the classroom echo responses scientists, engineers, and others who work in technical contexts give a question about what makes writing successful in their workplace:

> "*Clarity* is the most important element."

> "In priority order: *Clarity*, objectivity, brevity."

> " 'Good' written communication should be *clear*, relevant, and tone-appropriate (according to the audience)."

> "Successful written communication must be *clear* and unambiguous."

> "Again, I would say that *clarity* is critical."

While all five of these quotes are comments by real people that I collected about writing and communication in scientific and technical contexts, they need not be—I could just as easily have made them up—and you probably could have too. That is how conventional each statement is to our thinking about what makes communication successful. Each statement is so conventional and so uncontroversial that it could have been lifted from nearly any textbook, handbook, high school or college writing classroom, or casual discussion about what makes for successful writing.

As a previously trained creative writer and as a current language and writing scholar, I have long been fascinated by the word *clarity*. Any word that we put so much faith in and that does so much cultural work must have a really interesting story. In fact, *clarity* is not so much a word as a concept that as a culture we have bet everything on when it comes to explaining what makes for successful communication—the word is everywhere where there is explanation and discussion about what makes writing and communication successful. While lots of people would not claim to know much about writing and certainly not claim to be very good at it, nearly all people would at least know that *clarity* is the gold standard to aim for. If you don't believe me, start asking your family, friends, and coworkers what makes for successful writing or communication. I wager that some form of the concept of clarity will come up quite often. However, while the ubiquity of the concept is easy to document, understanding its logic and origins is much harder: Why this word and concept of clarity? Where did it come from? What does it really mean? Why is it so easy to agree on its centrality and importance?

The term and concept of clarity has on its own actually nothing to do directly with writing or speaking. Clarity is a characteristic of a material, such as water or glass, that we can see through—clean window panes are clear and easy and pleasurable to look through. Clear water is clean and essential to life. Diamonds have clarity. Clarity has to do with light and seeing or with a medium, such as air, which might be clear so that we can see the mountain vista in the distance, or polluted and therefore not clear at all. Clear air is also healthy air, and clear window panes make seeing through them easier and more satisfying, and clearer diamonds are more valuable, so in general clarity has a positive vibe and has a high value. Metaphorically, then, "clear" writing and communication is a window, the atmosphere after rain, or a diamond: it is effortless, healthy, and valuable. It gets out of the way and does not interfere with what lies behind it or the essential medium that constitutes it (language, atmospheric gasses, or carbon). Given all of these positive associations, or entailments, it's no wonder that we would want to associate ourselves with such a happy term and concept. How could we possibly go wrong with *clarity*? It is an absolute conceptual safe haven from the difficult and frustrating struggle that writing and communication demands of most people. Writing and speaking clearly can require herculean effort, yet the goal itself remains virtuous, ineffable and infallible. We could even say that *clarity* is doctrine for talking about writing. It is that important and powerful.

What is less apparent is how the metaphor of clarity is also a source of harm. In short, if language must be as clear as a sparklingly clean window, then we are in effect suggesting that we must do everything we can to ensure that language gets out of the way of what is really important, the message being conveyed. It's not hard to see how this value has two powerful outcomes: (1) that correctness (cleaning up writing) is essential to credible, successful writing, and, (2) that we don't believe that language actually shapes the message, or the knowledge; rather, we believe it is just a neutral purveyor of it. This chapter has already made the case for how both of these values cause harm. I'll explain further in chapter 3 how *clarity* functions linguistically and as a conceptual metaphor[31] that structures our tacit models for how communication works, which are part of what comprise the Communication Metaphor. These models for communication, which are linguistically, rhetorically, historically, and ideologically shaped and perpetuated, have caused a lot of harm, as the previous three anecdotes have started to illustrate and the rest of this book will develop in more detail. But here I'll offer the short writing class version of how to start to peel back the veil of virtue and comfort that shrouds the true consequences of *clarity*.

On the first day of a professional and technical writing class when I ask students to tell me about what makes writing successful, their comments generally follow a familiar logic: communication=clear=successful. In the follow-up discussion, I first assure students they are absolutely right that clarity is essential to successful communication—their education up to this point and their personal experience have not led them astray in that sense. For the most part, clarity, when achieved, works. The rub is in how to achieve it. I then tell them that, second, clarity is not, contrary to their expectations of what they'll learn in a technical writing course, achieved via the clean execution of well-polished prose that adheres perfectly to the rules of standard grammar and correctness; rather, clarity is a measure of how successfully the audience for your letter, report, or slide presentation has been able to understand your message and to respond appropriately. Clarity, therefore, is assessed by the audience for the writing or speech, not the author, the teacher (unless the teacher is the audience, which they so often are in school writing), or a grammar text or style guide. In other words, a perfectly grammatical sentence carefully edited for conciseness and the correct use of commas may still be clear as mud (metaphor alert) to its intended audience. Unfortunately, achieving clarity is a messy and complex process.

When clarity is understood as a prized quality of communication that is awarded not by adhering to the conventions and standards of correctness and style of Standard American English or White English Vernacular but by asking the audience if they "got the message," we change our understanding of how as writers we should go about making choices that are successful for our audiences. No longer can we assume that textbook or handbook guidelines for "clear" writing are the most reliable source of information on writing because clearly such resources are simply shortcuts to serving an exclusively SAE or WEV fluent audience. Instead, we should go directly to our audiences and ask them what they want or need to facilitate clear understanding.

This is especially true when audiences include those for whom SAE or WEV are not what they speak and write at home, and often at work as well. These audiences will not find writing or speech adhering to the conventions of SAE or WEV "clear"—in fact, likely the opposite. For any public writing, or writing that has an audience intended to include more than a small, known circle of people with a shared, well-defined aim and purpose, substantive audience research is essential, even though not always done, and, as developed in chapter 5, not always possible or ethical. In fact, adherence to the principles of "clear" writing of SAE or WEV have been used purposefully to exclude and alienate audiences, as was alluded to in the anecdote above about how violence against Native Americans has historically been perpetrated in part via treaties and compacts written in an English legalese that excluded the tribes from full participation in the negotiation or agency over the terms of the agreements. While of course not all common writing and communication situations in science and industry have the potential to do the same level of historical violence as the treaties the US government wielded as weapons against Native American interests, I hope that the allusion here to this history of violence at least serves as a shortcut to demonstrate how much is at stake when it comes to reexamining our most dearly and deeply held assumptions and convictions that are revealed when we examine how we talk about writing and communication.

Conclusion: Why the "Communication Metaphor?"

These four illustrations of the Communication Metaphor have not yet done much to define it, but they do begin to show the harm that it causes, every day, right under our noses in most English-speaking, primarily North American, technical writing classrooms and technical workplaces. One of the central arguments of this book is that the Communication Metaphor is in part to blame for this harm—both directly and indirectly. And while the Communication Metaphor is not unique to writing and communication in scientific and technical contexts, the bar is even higher for us in terms of revealing the harm that it does and how it does it. It is easy to point to how hot-headed political rhetoric purposefully uses language to exclude and to cause harm—even to incite physical violence—but it is much harder to see how technical and scientific writing and communication have the potential to do equally devastating harm, despite, and, in fact, *because* of, its mantle of clear-eyed objectivity and rationalism.

It may not come as a surprise that, as I develop my argument about the power of the Communication Metaphor and the harm that it causes and perpetuates, much depends on how we understand the key word *communication*. It depends on our individual and commonly held conceptions or theories about what communication is, how it works, and what makes it successful. Since we are all communicators in our daily lives that means that we all have a working model of communication that shapes our daily experience, even if we have never thought about it all that carefully. Since as a broader culture (primarily English-speaking North American) we value communication as essential to success, we need to understand what we mean by "communication."

This is a good point to shed some light on why I have chosen "Communication Metaphor" as the moniker for the complex of tacit assumptions, conventional practices, and historically rooted ideologies for writing and communication in technical contexts. There are multiple reasons for my choice, reasons that are definitionally justified and reasons that are both opportunistic and fun. First, "metaphor" is broadly recognized as an interesting and accessible complex language tool across all types of writing. The dictionary definition will be familiar to most readers: "A figure of speech in which a word or phrase literally denoting one kind of object or idea is used in place of another to suggest a likeness or analogy between them (as in drowning in money)."[32]

Examples of general metaphorical language:

You're a peach!

Time flies.

Blanket of snow

Examples from science:

Plant cells[33]

Imagine diving into a collapsing soufflé of intergalactic gas[34]

To say that I am going to write about metaphor creates a common ground for many readers because it connects my topic about technical and professional writing and communication to experiences you have already had in writing, English or language arts classes analyzing metaphor in primarily literary texts, although scientific texts are also full of metaphor. The notion of metaphor generally raises the issue that something interesting is going on with language and that it has to do with things not being quite what they seem. It is the power of metaphor to leverage new meanings by substituting one object or idea for another that underwrites the construct of Communication Metaphor. Simply put, I argue that the construct of the Communication Metaphor—which for now can be boiled down to how and why we find certainty in the statement "Successful technical communication is clear, concise and objective"—is a substitute for the rather messy reality that communication is one of the most challenging, frustrating, and highest-stakes activities that humans have evolved to engage in.

The fact that our broad agreements about what makes for successful technical communication are so simple and uniformly expressed suggests that these agreements are *being used in place of* something else more conceptually and practically complicated. This is hard to see because the Communication Metaphor has been so ubiquitous for so long for the sake of our convenience and has so seamlessly maintained the interests of power, whether that power is vested in individuals, companies and institutions, sectors of the economy or governments, or the West as a whole. Certainly, disturbing the equilibrium of the Communication Metaphor comes with risks. The something else that it stands in for would likely be controversial, at best, and might cause conflict, at worst, if admitted to daily conversation, the classroom, the laboratory, or the boardroom. Chapter 6 makes efforts to imagine what *the something else* might look like.

Underlying the argument of this book is my own belief that our currently widely held assumptions about how communication works and what makes it successful are first, harmful, and second, and more hopefully, *not the only ones possible*. My belief has been formed by both my experience as a technical communication teacher and my scholarly inquiry into what has sustained the Communication Metaphor over the last approximately 150 years. The tightly wound and deeply ingrained logic that professional success in America is linked to success as a communicator and that success as a communicator depends on conformity to so-called standard language practices and cultural practices that have been defined by and maintained by a dominant culture is, as I see it, evidence of the Communication Metaphor at work rather than the only possible conclusion. That is, I argue, there is an alternative set of assumptions about what communication is, its purpose, and how it works that would cause less harm of the kind illustrated by the scenarios above. Certainly, getting to a world where an alternative set of assumptions about writing and communication structures our daily language practices would likely involve a great deal of cultural change, much of it uncomfortable for many people.

This book has a lot of work to do to define the Communication Metaphor, to demonstrate that it is neither inevitable nor natural and that an alternative is possible. This argument will unfold across six chapters. After this first chapter, each chapter forwards the argument by responding to a fundamental question about the Communication Metaphor:

Chapter 2: How do we know that the Communication Metaphor exists?
Chapter 3: What does the Communication Metaphor mean?
Chapter 4: Where did the Communication Metaphor come from?
Chapter 5: How is the Communication Metaphor perpetuated and maintained?
Chapter 6: What is an alternative to the Communication Metaphor?

In addition, the logic of this book is organized to reflect a framework for social justice action within technical communication contexts. Scholars and practitioners in technical communication increasingly recognize that critique of technical writing and communication via a lens of social justice must be turned into practical action, and specifically coalitional action,[35] since the impact that any one individual can make is limited or can come at a high level of personal cost. It can't be overstated how important coalitional action is when working for any kind of change, and for social change in particular. Just as no person bears the whole burden of systemic injustice, no person can make scalable, systemic, and lasting change alone. If there is a way forward, it is collaborative, relationship-building, and mutually beneficial. The chapters of this book follow a four-step heuristic—or framework—for practical action that can address injustice:[36]

I. **Recognize unjust practices and our own complicity in them**

   Chapter 1: Harm Is Being Done

II. **Reveal unjust practices as a call to action**

   Chapter 2: Making the Communication Metaphor Visible
   Chapter 3: What Does the Communication Metaphor Mean?
   Chapter 4: Where Does the Communication Metaphor Come From?

III. **Reject unjust practices and opportunities to perpetuate them**

   Chapter 5: How Is the Communication Metaphor Perpetuated and Maintained?

IV. **Replace unjust practices**

   Chapter 6: Experiments in Imagining a Post–Communication Metaphor World

In this book, I hope to make progress in examining the assumptions and beliefs that have structured our understanding of what writing and communication in technical fields and workplaces is and how it works, with the aim to open up the possibility that an alternative may be possible. In other words, I hope to help end white silence about the harm that mainstream communication practices do. I aim to launch the Communication Metaphor as a meaningful term and referent for the beliefs, assumptions, and practices that structure our daily experiences of writing and speaking, especially in technical contexts.

Chapter 2

Making the Communication Metaphor Visible

When I teach a course on introductory technical writing, on the first day I often present pictures of two Sumerian clay tablets (circa 3300 BCE–3000 BCE[1]) from the collection of the British Museum. The small brown tablets, only about eight centimeters by five centimeters, display pictographs that record quantities of beer and barley. What's striking about the tablets is not the use of cuneiform, per se, but the tablets' communicative purpose. These tablets are evidence that in the cities of Sumer a system had been developed by the rulers and officials to monitor and control economic activity. Writing began as a visual system for records of the production, storage, and transfer of commodities, and for establishing ownership, trade obligations, and tax payments.[2] This is to say, I tell my students, that the very first writing was technical writing. As early cities and states organized into bureaucracies that could monitor and control growing economies and populations, the means to tabulate and record evolved along with them. Civilization, I tell my students, simply isn't scalable—or arguably possible at all—without technical writing.

The idea that technical writing—normally overlooked as boring, mostly benign, and, at best, a secondary, background concern to the real work and activities of the scientific and technical world—is one of the foundational inventions of civilization[3] really catches the students' attention. And while it is easy to overstate,[4] and I am shameless in doing so given my own passion for seeing the world spinning around an axis of technical writing, this grand idea serves an important purpose: it suddenly makes visible and elevates the importance of many types of writing that have heretofore been primarily invisible: grocery receipts, paycheck stubs, grade reports, student financial aid applications, electric bills, car insurance policies, instructions for operating a new washing machine, terms of service agreements, lab notes, and technical reports. In other words, technical and professional writing suffer from a visibility problem—most of our lives we simply haven't noticed how essential it is that our world is saturated in written documentation, now largely digital. And this is a problem because it is hard to value and have a deep understanding of things that we can't see or that we don't take much notice of.

The aim of this chapter is to show evidence that the Communication Metaphor is simultaneously invisible and ubiquitous as it exists in everyday and conventional language use. In the first chapter I argued for how our conventional assumptions and practices about scientific and technical writing and communication cause harm, even if we are not aware of it. I lay the blame on the Communication Metaphor, which I referred to in a number of ways, including conventional language practices regarding correctness and credibility and how those exclude dialects, ways of thinking and doing, and types of knowledge—and, most importantly and potently, people. The Communication Metaphor, I argued, is a *complex* of tacit assumptions and practices for technical and scientific writing that are widely considered standard and normal to what counts as successful written and spoken communication in technical contexts. The problem is, however, that these assumptions and language practices are largely invisible to us as we use them in our daily work with hardly a second thought. The first task then, is to make the Communication Metaphor visible so that we can concretely and materially understand (and believe) that it exists. In this chapter I'll start to unpack how we notice the Communication Metaphor, which will start to give us clues about how it functions in the language we use to talk about writing and communication. In this chapter we'll discuss aspects of the Communication Metaphor that make it visible to us so that we can then go on to discuss what it means in chapter 3.

Before going forward with the project of making the Communication Metaphor visible, I want to say a little bit more about what I mean when I say that the Communication Metaphor is a *complex* of tacit assumptions about and practices for language use. As I sat at my desk working on an initial draft of this chapter in the middle of a hot summer in the Pacific Northwest, the word *complex* had a special resonance. This year was shaping up to be the most historically devastating wildfire season for the western United States and Canada, and it was scary—hundreds of thousands of acres of tinder-dry forest and wildlands were burning and largely overwhelming the fire crews assigned to contain them. Wildfire smoke consumed the air of much of the western United States and Canada—we were living in the environmental apocalypse wrought by climate change.

Even worse, there were so many individual fires that they were merging together and becoming a singular wildfire *complex*—just like our many assumptions about how language works and how we use it can also merge together into a single concept, or complex, the Communication Metaphor.

The USDA Fire Service defines a complex as "two or more individual incidents located in the same general area which are assigned to a single incident commander or unified command."[5] This definition is useful for

Figure 2.1. Like a complex of multiple forest fires, the Communication Metaphor is comprised of many separately ignited language problems that are best fought as a single, centrally managed conflagration.

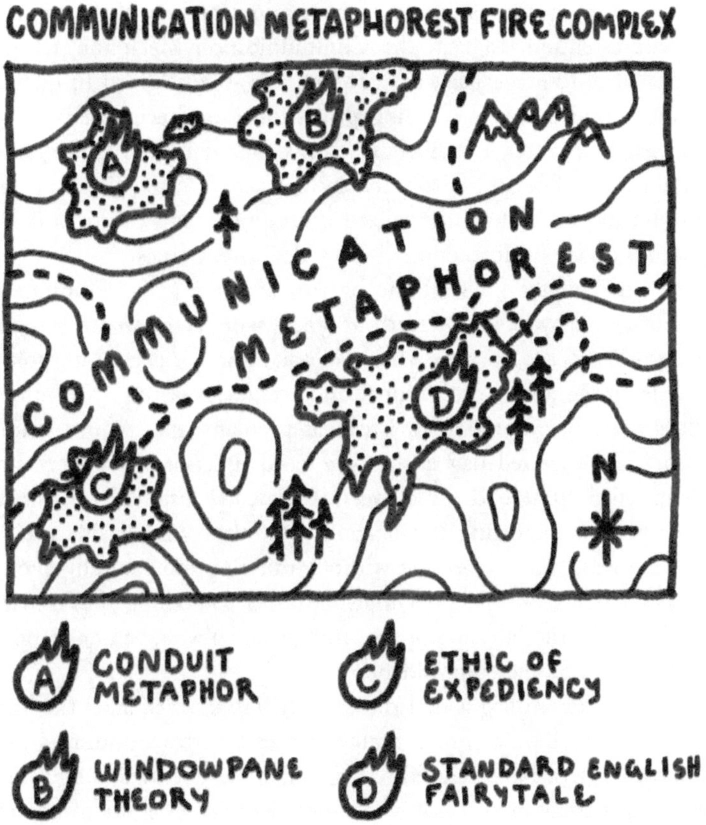
COMMUNICATION METAPHOREST FIRE COMPLEX
A
B
C
D
COMMUNICATION
METAPHOREST
N
A CONDUIT METAPHOR
B WINDOWPANE THEORY
C ETHIC OF EXPEDIENCY
D STANDARD ENGLISH FAIRYTALE

thinking about the Communication Metaphor as separately ignited incidents (e.g., values of correctness in standard technical and professional communication textbooks, historical changes in what the word *communication* means, Wikipedia requirements that police what counts as knowledge, etc.) that have merged together into a centrally managed conflagration emitting smoke that obscures our views, chokes our respiration, and forces people from their homes. At the end of this chapter, we'll map the *complex* of the Communication Metaphor to make visible the multiple assumptions, values, and practices that shape how scientists, engineers, and technical professionals understand technical writing.

However, what we assume about how communication works and how we talk about it is actually not the same as a forest fire. A forest fire imposes itself upon our senses of sight (flames, ash, smoke plume), touch (heat), and smell (smoke) such that we are forced to react immediately to preserve the safety of our bodies and our homes. The Communication Metaphor does not normally impose itself in such a visceral way that demands our immediate attention. In fact, the Communication Metaphor is not really visible at all during everyday life because it is so embedded in our tacit or fully normalized practices of using professional and technical language. As I discussed in chapter 1, while the harms caused by the Communication Metaphor are visible and deeply felt by the people most directly impacted, the dominant and widely normalized assumptions and practices that cause this harm go largely unrecognized on a day-to-day basis.

In this chapter I'll start the project of making the Communication Metaphor visible by looking directly at the keyword *communication*. The word *communication* raises red flags when we look at how it is used in spoken and written language about professional and technical writing. Metaphorically, red flags draw our attention to issues or problems that require immediate attention. The first red flag is that the word functions as a black box, or, as a simplified front end for a messy back end. Just like we use laptops everyday without knowing much about how they work inside their cases, we use the word *communication* without thinking too carefully about how it works. A black box is often invisible until something breaks (e.g., the motherboard in the laptop stops working), and then opening it up to see what is inside becomes unavoidable.

The second red flag is that the word *communication* takes the linguistic form of a nominalization, or a noun form of a verb (communication [n.]; communicate [v.]). Nominalized terms, which are essentially linguistic black boxes, are generally excoriated in technical writing style guides as bad style because they introduce unnecessary abstraction, wordiness, and a pompous tone to sentences—all faux pas of conventional style. Aside from being bad style, however, nominalizations also perform a special and powerful function of language: they have the power to transform a concrete action within a specific context (an active verb) into a capacious idea (an abstract noun) that can take on a cultural life of its own. Whether as a matter of style or as a matter of the power of how language works to create and reflect cultural mores, a commonly used nominalized term, such as *communication,* is one we should pay extra attention to.

Once visible, we can see how a black-boxed, nominalized term such as *communication* carries a lot of freight for us in daily conversation and writing. Not knowing more about what is behind it carries the risk that we are doing harm without being aware of it. While it's easy and breezy to say or write that "Good communication skills are very important for scientists and engineers," it is much more difficult to actually say what *communication* is, how it works, and what makes it good—and thank goodness we don't have to explain all of that every time we write or say the word! Important to unpacking the black box of *communication* is to look at how this important keyword word is used in context and how its use and meaning has evolved over time into the modern understanding we have today. Doing this makes visible how the meaning of *communication* changes to reflect the values and ideologies of a given historical era.

Finally, at the end of this chapter I'll map the *complex* of the Communication Metaphor to make visible the multiple assumptions, values, and practices that shape how scientists, engineers, and technical professionals understand technical writing. What I am calling a map is actually the "coding" of responses that scientists, engineers, and technical professionals wrote in response to a survey about professional and technical writing and their careers. Coding is the qualitative data analysis method process of systematically looking for meaningful themes in the content and language structure of textual data. Coding the survey responses makes visible how the multiple aspects of the Communication Metaphor are present in the survey responses.

Communication as a Black Box

A research result that I might report to my colleagues could go like this: 100 percent of scientists, engineers, and other technical workers surveyed answered "Yes" to the question "Is writing and communication important to your work?[6]" The fact that there is 100 percent agreement about the importance of writing and communication is not, however, the significant research finding; rather, it's the unanimity regarding the topic of *writing and communication*. If there is broad agreement about the value of writing and communication, then it's reasonable to wager that there must be a common understanding about what they are and how they work. The word *communication*, which for right now we will assume to be an umbrella category that includes writing as but one mode, is not defined in this question; as a result, survey respondents were required to supply their own understanding of what *communication* means to formulate their response. There is no definition because for most people reading this question a definition wouldn't be necessary to give a "yes" answer. Instead, the uniformity of positive responses reflects an already widely held value and understanding of *communication* among people in technical professions. To take this line of thinking a step further, we might also say that we have been trained via schooling and experience to accept a broadly shared understanding of what *communication* means with the outcome that research results like this one are unsurprising. However, the content of this broadly shared understanding remains largely invisible, as it is in this survey question, in everyday use of technical language.

One way to unpack the significance of our blindness to the term *communication* is to understand it as a *black box*. A *black box* is a simple interface that reduces the complexity of a system so that it can be used without a full knowledge of how and why it works.[7] The concept of a black box as a conceptual tool for revealing hidden complexities originated in French philosopher, sociologist, and science and technology studies scholar Bruno Latour's work to explain how widely accepted scientific facts stand in for the complex scientific uncertainties and controversies that they emerge from. Latour pins the origins of the term to cybernetics. Cyberneticians draw a box to represent a complex machine when it is the input and the output from the machine that they care most about, rather than the inner workings of the machine. In figure 1, the boxes for "Sensor," "Controller," and "Other system," stand in for the complex technologies that they represent, or black box.

Latour argues that the concept of black boxes can be extended to scientific concepts as well, such as the double-helix shape of DNA. While for Watson and Crick the shape of DNA was the outcome of a "fierce challenge"[8] to overcome the controversies and challenges that shaped their

Figure 2.2. An illustration of a cybernetic loop diagram representing a cycle of causality or feedback in a complex system.

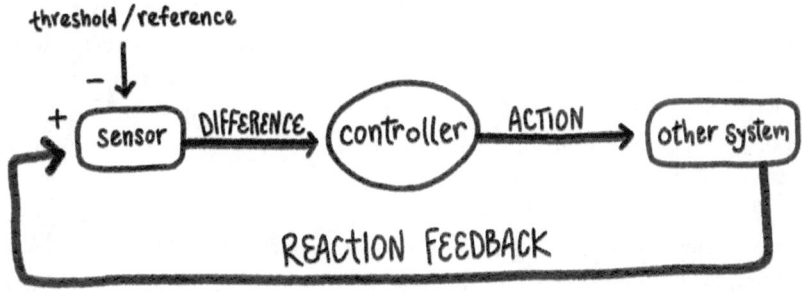

model, which ultimately won them a Nobel Prize, for researchers after them the double-helix was a "basic dogma" that they could incorporate into their research without having to reiterate the controversies that shaped Watson and Crick's work. In short, a black box literally draws a line around the controversial concept to contain its inherent complexity.

When controversy is black-boxed, a concept can be repurposed at a higher level of abstraction—for better and for worse. Latour argues that black boxes can be material and technological, such as the iphones, laptops, and Amazon.com webpage that we rely on in our daily lives without having to fully understand how they work. Black boxes can also be conceptual, in the sense that the double-helix is now a widely accepted concept even though the scientific controversies at its origins are no longer cited every time that the structure of DNA is mentioned. Likewise, I am arguing here that we can extend this concept of the black box to the word *communication*. *Communication*, I argue, functions as a black box, or as an uncontroversial interface for a deep well of uncertainty and controversy over how communication works and what it means.

Black boxes are actually very useful to us and serve an essential function in daily life. For example, our phones and computers are simplified interfaces that enable us to perform actions that require the existence of a great deal of technological and organizational complexity, most of which we don't have to think about on a minute-by-minute basis in order to go about our daily business. This morning when I made coffee, I didn't have to think about engineering a coffee maker or locating and harnessing a source of electricity, I just added ground coffee and water and pushed a button.

Likewise, the word *communication* enables us to function smoothly in our lives as technical professionals. We accept that it stands in for a process that is essential to achieving success in our careers. In short, if we input widely valued common language practices, such as thinking about the level of technical expertise of the audience and making smart language choices to write clearly, the output will be a published paper, an approved technical report, or a promotion. This creates a positive feedback loop that sustains the system over time.

This is all to say that black boxes are in general good and necessary things—they are all around us and are very helpful in making a complex, technological, and heavily bureaucratic world navigable. When something fails, however, that is when we need to look inside the black box and understand and become accountable for the complexity that the black box normally obscures. Part of the purpose of this book is to argue for how the black-boxing of the word *communication* is evidence that the Communication Metaphor is failing, given all of the harm that it causes, and that our response should be to unveil the messiness at its roots.

With this new understanding, let's look at a few more examples of statements from academic journals and technical writing textbooks that include *communication* in use as a black-boxed term:

> [1] Excellence Theory suggests **communication is valuable** to an organization since **communication** leads to strategic relationships with the public.[9]

Figure 2.3. An illustration representing communication as a black box sustained by a feedback loop of input and output.

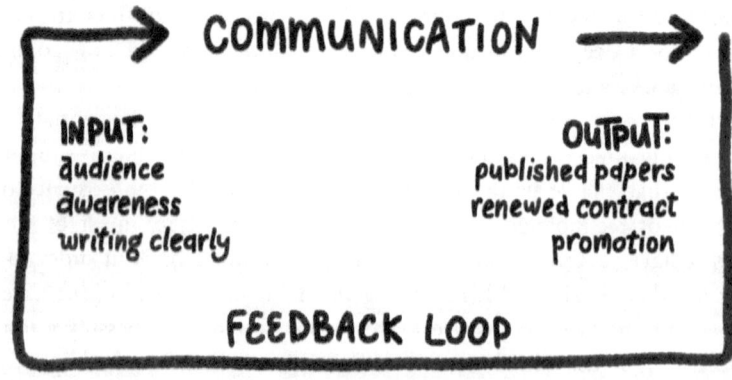

[2] Effective **communication is essential** to the success of implementing a discovery service. Implementation teams need **to communicate** with multiple audiences both outside the organization (with vendors, product user groups, and other libraries) and within the organization (with staff, users, and other stakeholders).[10]

[3] **Communication is essential** in the scientific community and in STEM careers.[11]

[4] An important aspect of home-school **communication** is providing clear **communication** . . .[12]

Each of these examples states a high value for communication without saying much, if anything, about what it means because the complexity is hidden behind the simple interface of a familiar and common term. We know this empirically because no definition of the term is provided for the reader in the sentence, or in the context of use, even though the term carries the full freight of the meaning in these statements. In other words, from these statements alone it is not possible to discern what "communicate" or "communication" actually mean or how the process works. You might suggest that the definitions could appear elsewhere in the texts that these statements have been lifted from. However, since it isn't necessary to have those definitions in front of us to agree that these statements are true or have value, it is likely that the text surrounding these statements also did not explicitly define what "communication" means (and I also know so since I checked in each case!).

The ubiquity of the use of the words *communicate* and *communication* as black boxes in the context of professional and technical writing is further evidence that the Communication Metaphor for writing is one that functions pervasively in our language use separate from conscious thought. What *communicate* actually means, the authors of examples 1 through 4 rightly assumed would be tacitly supplied by the reader of the sentence. The authors could reliably expect readers to fill in a stable, common definition of their own that the reader had already learned externally to the text at hand, most likely from schooling, personal experience, or the culture at large. And how handy this is, since once we start to think about it (and writing, rhetoric, and communication scholars have done so extensively in the scholarly literature), we quickly realize that it isn't at all clear what communication is or how it works. Recognizing the term *communication* and its other forms (*communicate, communicator*) as black boxes that conceal a great deal of complexity behind the user-friendly interface of an uncontroversial common term is the first step toward acknowledging how much work we have to do to fully unveil the Communication Metaphor and to begin mitigating the harms that it causes.

Definition by Repeating Black Boxes

Another clue that the term *communication* functions as a frontispiece for a long and complicated—and most notably missing—treatise on what it means is how often it is defined by tautology, or by definition using the word that is being defined. Looking again at examples of communication as a black-boxed term:

> [1] Excellence Theory suggests **communication is valuable** to an organization since **communication** leads to strategic relationships with the public.[13]

> [4] An important aspect of home-school **communication** is providing clear **communication** . . .[14]

You'll notice that the sentences actually form tautologies, where what *communication* means is developed later in the sentence by the same word, or its verb form (*to communicate*), without adding any new information about what *communication* means or how it works. Here is another example of the communication tautology at work from a one of the most commonly assigned standard technical communication textbooks:

> [1] The purpose of **Technical Communication** is to help you learn the skills you need to **communicate** more effectively and efficiently in your professional life.[15]

This textbook, even at 733 pages long, does not ever explicitly theorize or provide a model for how communication works for its readers, relying entirely on readers to supply their own conceptual understanding. This is not to say that this text is not fairly sophisticated and thorough in what it has to teach technical writing students. Students learn to approach technical communication as a problem-solving process; they learn that drafting technical and workplace documents is a multistage process, and that they must very carefully consider who their audience is when making rhetorical choices, such as about genre, how to best structure information, and word choice. You might ask what is wrong with teaching and learning these topics about technical communication, and, of course, the answer is nothing at all—or at least very little—especially from a strictly pragmatic point of view. Yet, when we view instances of the communication tautology as evidence for how much we have stopped thinking about how communication works, and therefore how much harm our assumptions and tacit practices might cause, this omission takes on a greater weight.

As technical communication teachers, students, or practitioners, we are actually all very familiar with the logic of the communication tautology that structures example 5 and have likely observed or engaged with it often. Consider this series of conventional statements that paraphrase many opening chapters of standard technical writing textbooks or that outline a conversation with a new employee, research assistant, or mentee:

> Why should I spend time learning about technical **communication**?
> Because **communication** *is important!*
>
> What is **communication**?
> *Writing and speaking* **to communicate** *information and ideas to various audiences.*
>
> How does **communication** work?
> *By writing and speaking well so that others can understand what you want* **to communicate.**
>
> Why is **technical communication** important?
> *Because* **communication is key** *to success in your career.*

While this series of statements oversimplify how most people would actually explain how they understand writing and communication in the context of their lives and professions, my purpose here is to make visible how handily the terms *communication* and *communicate* stand in for any need to explain or theorize the mechanism behind how they actually work. Given the example of the standard technical writing textbooks and your reflection on your own experiences that resonate with the statements above, it's fair to say that explanation by tautology actually seems to be sufficient most of the time. So, what is the problem?

While we all have a largely unconscious working model of how communication works that informs our everyday actions when it comes to writing and speaking in professional and technical contexts, there are few situations in our daily work in which discussing this working model explicitly is actually necessary to get the job done. Part of what maintains the Communication Metaphor as a powerful and ossified element of a dominant culture that White speakers of English have historically preferred as "normal" or "correct," or treated as intrinsically superior to other forms of culture and English language, is the fact that, in balance, it seems to work. In other words, it serves many people and institutions most of the time—except for those that it doesn't.

Communication as a Nominalization

Conceptualizing the word *communication* as a black box helps us to see it as something that requires excavation to explore its full, hidden complexity, or as a red flag for a problem that demands more attention. A linguist, however, might have been thinking about another language tool for hiding things while reading the last section. They might be wondering about the word *communication* as a linguistic black box, or nominalization. A nominalization is a noun form (*communication*) of a verb (*to communicate* / *to commune*). Like the concept of the black box, nominalization is another powerful analytical tool to help us diagnose the Communication Metaphor. Here we see what *communication* as a nominalization looks like in use in the words of a research scientist:

> [1] **Communication [noun]** is everything! If you can't **communicate [verb]** your interesting research findings, they are pretty useless.[16]

There, of course, isn't really anything exceptional about the structure of these two sentences or the use of the word *communication* in the noun form followed in the second statement by *communicate* in the verb form. We are used to the nominalized form of the word—on its own it is nothing exceptional. You might also notice that example 1 is also another example of the black-boxed tautological logic discussed above—where the meaning of the nominalized form is explained via the repetition of the word in a follow-up phrase or sentence.

In this section I'm going to discuss what it means to look at the word *communication* as a nominalization in light of what linguists know about how nominalization works in sentences, stylistically and conceptually. Understanding the word *communication* as a nominalization tarnishes its sheen as a word often experienced as, at best, a feel-good go-to word that we assume reliably creates a shared understanding and experience (except for whom it doesn't) and, at worst, a neutral, uncontroversial term in need of disturbance. While nominalization is not a characteristic unique to the word *communication*, the fact that *communication* is a nominalization gives us even further reason to recognize it as a red flag for a deeper problem.

NOMINALIZATION AS AN ISSUE OF STYLE

Nominalization is most commonly taught as a matter of writing style. You might already be familiar with the fact that most professional and technical writing textbooks and handbooks consider nominalization to be a case of poor writing style that should be avoided, offering advice such as this:

> A nominalization is a noun form of a verb that is often combined with vague and general (or "weak") verbs like make, do, give, perform, and provide . . .
>
> [For example:] The staff should perform an **evaluation of** [**evaluate**] the new software.
>
> Avoid nominalizations when you can use specific verbs that communicate the same idea more directly and concisely.[17]

. . . and end with a severe, morally inflected warning to writers:

> If you use nominalizations solely to make your writing sound more formal, the result will be affectation.

Even if technical writers have no intention to write in an artificial style with the explicit intention to impress, the style guide's warning does come for good reason. Stylistically, the stronger version of the "evaluate" sentence would be: "The staff should evaluate the new software." The alternative sentence has better style because it has fewer words (is more concise) and so is quicker to read and takes up less space. The verb *evaluate* is also less abstract than "to perform an evaluation" and so is quicker to comprehend, especially for non-native English speakers.

A second example[18] shows how editing a sentence to replace passive voice ("to be" verbs) with active voice (action verbs) also minimizes the use of abstract, nominalized terms (in bold):

> Unedited: There were **expectations** by the laboratory safety committee that its protocol **submission** would meet the deadline.
>
> Edited: The laboratory safety committee **expected to submit** its protocols by the deadline.

The edited version is shorter (is more concise), more concrete, and more direct, as well as sounding an awful lot less pompous! From a stylistic point of view, it's a winner.

Nominalization, however, is more than just a style matter to avoid. It also has, like all parts of language, a purpose, or a function, for "how people use language with each other to accomplish everyday social life."[19] When we talk about the function of a part of language, we are interested in more than the dictionary meaning of the word or identifying the grammatical elements of the sentence. We are interested in how the *form* and *structure* of the word or sentence part also contributes to meaning. In the case of *communication,* we are interested in the powerful language structure of nominalization that can pack what is a messy, complex process into a single, palatable word. The invisibility trick of nominalization is to take a big, hairy complex and make it into a more streamlined, comfortable thing, or to "encode processes into nouns."[20] It's this trick that raises the red flag on *communication.*

Nominalization as Encoding Complex Processes into Single Terms

One of the tricks of nominalization enables the transmutation (nominalization alert!) of specific events and processes that happen in time into abstract concepts that can be generally referred to and transported across texts. In both historical and contemporary scientific writing this can be useful and serve the purpose of creating new scientific knowledge. This excerpt from an astronomy text from 1777 displays how the linguistic process of converting a new event that happens in time into an abstract process stripped of its background information can happen within a single paragraph.[21]

> The orbit of every planet is in a plane paſſing through the ſun, **which planes are inclined to one another:** thus in fig. 4. let ABCD repreſent the earth's orbit, or plane of the ecliptic; this is taken for a ſtandard, from which the inclination of each orbit of the planets, as EDFB, is meaſured. **The inclination** of the orbit of Mercury is 6°,52' that of Venus 3°,33', of Mars 1°,52', of Jupiter 1°,30', and of Saturn 2°,30'.[22]

The conversion of new information ("which planes are inclined to one another") into backgrounded information for the new sentence subject ("inclination") via nominalization is a normal feature of scientific writing. More than just a matter of bad style, nominalization is fundamental to the grammar of scientific English because it accommodates the nature of scientific knowledge as abstract and generalizable.[23]

Linguist Iria Bello argues that this type of nominalization also serves a cognitive function.[24] Since the second part of the paragraph is built on the foundations of the first part, the text is able to efficiently build toward increasingly streamlined, abstract ideas that can be repurposed later in the text or applied outside of the text. The hazard of the compactness that nominalization provides in a text is the exclusion of readers who do not have access to the original background information and therefore may not be able to fully understand the idea. This effect is similar to how DNA became black-boxed such that the term *DNA* could simply stand in for a concept that was once a site of scientific controversy.

The power of nominalization to encode complex processes or ideas into single terms[25] also serves to transform a contextually bound, specific thing into an abstract entity with the salience of a cultural phenomenon. For example, an advertisement for plastic surgery announcing "Good looks can last you a lifetime" gives the abstract notion of "good looks," which is a nominalized form of the more concrete "You look good!," the status of an inherent state or property of living.[26] The transformation of the verb *look* into the noun *looks* creates the new notion of "good looks" as an ill-defined but apparently broadly known and desirable phenomenon that can be culturally manipulated. What does "good looks" really mean? That, of course, is up to the context and the individual; however, the statement assumes that any reader will accept that "good looks" exist. Nominalization is a powerful engine of what are essentially cultural memes!

Can we imagine how the same transformation applied to the word *communication* makes visible the same trick of turning a concrete, lived relational process (to communicate) into a cultural meme? On its own the single word *communication* offers no information about the process that it substitutes for, even if we accept that it stands in for the entire process of human (and animal, and machine) use of symbols, such as written language systems or marks, to share experiences, knowledge. and information, and to build relationships based on mutual understanding. In other words, *communication,* as a nominalization, stands in for the messy, complex, multifaceted, and difficult process that both academic and self-help book-long treatises have striven to demystify, and yet still fall short.

Nominalizations, such as *communication* or "good looks," serve a highly pragmatic function in language use—they enable us to efficiently pack into a single word what might otherwise require extensive explanation every time a writing context requires reference to the process. It's hard to imagine having to explain how we use symbols to create mutual understanding every single time the topic of communication arises. Yet, this same utility is a red flag that we should pay attention to. When a process is nominalized, the semantic, or meaning, content is stripped out, and the reader is left to supply the meaning of the word on their own, as is the case with the statement from the research scientist:

> [1] **Communication [noun]** is everything! If you can't **communicate [verb]** your interesting research findings, they are pretty useless.[27]

Few, if any, would disagree with this sentiment, despite the total lack of information about what *communication* or *communicate* mean in this sentence. Since the nominalization *communication* is a common word, there is low risk for the writer in offering a reader little to no information about what it means or how it works. Why is this? When it comes to the word *communication,* I argue, we have a broad cultural agreement that there is a common understanding of what it means. And it would seem that this is a safe assumption since the nominalization *communication* is so ubiquitous—until we take into account who and what these broad cultural agreements leave out, with the outcome being the harms detailed in chapter 1.

Communication as a Term in Historical Context

In the early twenty-first century, the word *communication* is visible, and audible, all around us in textbooks, lectures, writing and speaking handbooks, career advice, self-help books, and so on. The above section referred to it rather casually as a "cultural meme." It's a word that is key to how we talk about the value of speaking and writing in both interpersonal and professional relationships. As the previous section developed, the word *communication* is an abstract concept in the form of a nominalization—a single term that stands in for a complex process or domain of human experience that philosophy, communication, writing, and rhetoric scholars and everyday people can theorize deeply about, or that, conversely, many people don't really have to think much about at all. Whether we think about what it means and how it works or not, the word *communication* carries a big load as the key to success in the modern world. Interestingly, however, this has not always been the case. *Communication*, while not a modern word, is, as we broadly understand it today, a modern concept.

This chapter began with two Sumerian tablets circa 3300–3000 BCE that are evidence that the earliest writing (cuneiform) served the bureaucratic purposes of managing increasingly large and organized economies and populations. An interesting corollary to this principle is that the earliest writing was *not* for the purpose of interpersonal communication, whether for business or personal purposes. In fact, even the first alphabetic writing in the Greek world was short inscriptions on eighth-century vases, which were statements of ownership or names of deities or offerings to a god.[28] Evidence of the first alphabetic letter writing for the purpose of sending a personal message between people across a distance points to Persia around 500 BCE and the Greek letter-writing tradition. Letter writing in English for conducting business affairs didn't appear until the fifteenth century, although prior to that business affairs had been conducted throughout the Middle Ages in Latin and French by a specialized class of professional scribes working largely within the church.[29] Written interpersonal communication, whether to conduct personal or business affairs, couldn't be practiced by the merchant or professional classes until English became a shared vernacular and access to education and literacy became more broadly available.

That writing evolved to manage bureaucracies before it evolved to facilitate interpersonal communication is important because it makes us aware of how the earliest writing developed independently of what we understand as a modern notion of person-to-person or writer-to-audience written communication. In fact, it has been only since the late nineteenth century that the ability to communicate successfully with others has become a defining personal trait related to success in private and public life. The modern idea that our personal and business relationships will flourish in "a utopia where nothing is misunderstood, hearts are open, and expression is uninhibited"[30] is one of the characteristic concepts of the twentieth century.[31] It speaks to the centrality of communication in shaping our worldview in the early twenty-first century that we would look to *communication* as the source of both success and failure in some of the biggest challenges of our age. For example, in the personal realm we broadly accept that good communication is essential to making a marriage or life partnership work well. In the scientific and technical spheres, the cutting edge of research depends on coordinating massive multinational big science projects, such as the Large Hadron Collider at CERN in Switzerland,[32] which require extensive communication in the political, scientific, and economic realms that big science relies on for its existence. In other words, modern life depends on the concept, means, and abilities of successful communication.

To see how the use and meaning of the word *communication* has evolved significantly over the last two hundred or so years, we can look to evidence in large research databases of contemporary and historic texts—what linguists call corpora. Large corpora include all kinds of genres of fiction and nonfiction texts. A search through a corpus of historical nonfiction texts[33] (fiction was excluded from the search) turns up evidence that the notion of communication in technical writing has evolved from being a physical connection between distanced objects, places, and people created by rivers, waterways, or other types of physical means to, a century or so later, a fully abstract noun that refers to an idealized concept or process located in neither time nor space.

In nineteenth-century nonfiction texts, *communication* refers to a connection created by waterways or railroads or disrupted by mountains; for example:

> Why not imagine what is much more within the range of probability, that by some tremendous convulsion of nature, the Andes should be thrown into the ocean, and thus interrupt the *communication* we now have with Asia and South America?[34] (1827)

> At certain seasons a perfect water *communication* between the Mississippi and lake Michigan is formed through this river, canoes being able to pass easily from it into the Fox river of

Green Bay, but the navigation is interrupted in several places by falls.³⁵ (1822)

Communication can also be achieved via plant roots. An treatise on growing grape vines from 1837 frames communication in terms of roots forming a connection with the soil in order to grow:

> After the first of May, care must be taken to keep the soil round the cuttings constantly moist. For this purpose supply each cutting as often as required, according to the state of the weather, with about a pint of soap-suds; and continue so to do until it has formed a *communication* with the soil, which will soon be rendered apparent by the protrusion of a shoot . . .³⁶ (1837)

Jumping ahead in time, in the twentieth- and twenty-first-century, texts in the Corpus of Historical American English show that *communication* has been abstracted to a primarily interpersonal process for creating connections, rather than the connection itself:

> "I've learned so much about how important communication is, how to communicate how you really feel," he says. Communication is words, for sure, but also much more. (1994)

> Communication is especially important if there are cultural differences to consider.³⁷ (2003)

In addition, among our contemporaries in science and technical fields, communication has also been abstracted into a highly valued concept. This is a comment by a researcher of biophysics, but one that could be made by most readers of this book:

> Writing and communication are important to my work.³⁸ (2022)

What happened over these roughly two hundred years that would explain the shift in *communication* from the concrete context of waterways, roots, and other natural features to a lofty abstract concept focused on human interaction? Of course, a lot of things happened: there were three major wars: the Civil War, World War I, and World War II, all of which placed significant demands on communication as a tool for warfare. In addition, the economy shifted from a primarily agrarian economy to an industrial economy and then to a knowledge-based economy. In accordance with the economic shifts and the development of new technologies, there was an expansion in the areas of knowledge that supported them, such as management science, the social sciences, and psychology. In addition, the rise of mass literacy, the book publishing trade, and cultural changes that centered the development of the individual resulted in the availability of this new knowledge to the general public in the form of management, career advice, self-help, and parenting books. A notion of corporate or interpersonal *communication* was a foundational concept for all of these endeavors and became a characteristic concept of the twentieth century.[39]

As new technologies, or means, for communication were developed to meet the demands of warfare and a changing economy, the challenges central to achieving successful communication also evolved. During the time when creating communication, or connections, between natural features, communities, or people required passage in a boat, along a road via foot, horse, or carriage, or by air via a pigeon, flags, or smoke, the central problem was the type of terrain, the natural obstacles, and the physical limits of any given means of transportation. This is to say that up until the inventions of the telegraph (1830s) and the telephone (1870s), the concept of *communication* was essentially synonymous with transportation:

> There is no boat in the island, and the only *means of communication* between Wooding Island and Fernando is a <u>small raft or catamaran</u>, which is carefully kept in one of the forts, and is capable of hearing only two men.[40] (1815)

> Beside the main peninsula, the city comprises another peninsula, called South Boston, connected with the former by two free bridges; and the island of East Boston, with which *communication* is kept up by <u>steam ferry-boats</u>.[41] (1837)

However, with the development of technologies such as the telegraph (1831) and the wireless radio (1890s), and eventually digital technologies, the older notion of communication as transportation was retrofitted into a quasiphysical connection with a much more abstract notion of the physical obstacles to crossing space and time:[42]

> This is easy to understand, given that, as I've already said, *communication* is central to just about everything anyone does on the Net.[43] (1998)

While the "Net" is also made up of physical components of a computer network (computer terminals, servers, cable connections, etc.), communication in this sense doesn't refer to the hardware connections themselves but to an activity that is happening via those connections. A user of the internet can quite happily have no knowledge of how it works, unlike the knowledge of the terrain needed by a captain of a ship navigating a water passage or a Pony Express rider traversing between Missouri and California in 1860. In a sense the electronic age enabled the black-boxing of communication by removing the physical barriers to its success. If the method and the means of communication are mutually dependent, such as when a river requires a boat to traverse, then if the means of communication (the boat) becomes invisible, then so might the method (the river). That is, most of us don't need to know the terrain of the internet to traverse it or even know how email or instant messaging works, so both become invisible, or black-boxed, to us.

Without the challenges of traversing unforgiving landscapes and bad roads or waiting for pigeons, the problem of successful communication shifted away from transport to understandability. The invention of the telegraph in 1831 and then wireless radio in the 1890s and the revolution that it ushered in shifted the central challenge in communication from whether the message gets to its destination at all and in how much time to whether the message can be understood by the recipient as intended by the sender. The main problem shifted from one of shipping to one of decoding and interpretation. Along with changes in communication technology came new models, or metaphors, for explaining how communication, now understood primarily as a problem of understanding rather than a problem of transport, works. In chapter 3 we'll examine the transmission model and other metaphors that shape our assumptions about how writing and communication work in more detail. Making visible and understanding these metaphors that are embedded in conventional language about technical writing reveals a lot about our modern assumptions and conventional practices in the classroom and the technical workplace.

How Can We See the Communication Metaphor at Work?

The primary aim of this chapter has been making the Communication Metaphor visible so that we can believe that it exists. So far the chapter has focused on the word *communication* as a black-boxed, abstract nominalization that reflects the context of the modern world rather than a timeless value independent of the economic, technological, or political threads of an era.

It is time now to actually see what the Communication Metaphor looks like. We'll see how it shows up in the conventional language of scientists, engineers, and technical professionals.

In the preface I shared the answers scientists and technical professionals offered to the question about what makes writing and communication successful. The answers elicited comments that would come as a surprise to no one:

Successful written communication is:

- concise, accurate, elegant, informative, and engaging
- accurate, relevant, easy to understand, and accessible
- short and to the point, with the facts only, no opinions
- clear and concise wording to explain complex concepts
- clarity and brevity

The problem with these statements about writing is not that they aren't true for many people much of the time, it's that in their conventionality and broad acceptance, the values and ideologies that motivate them, have become invisible to most people most of the time. Chapters 1 and 2 have already developed aspects of the Communication Metaphor, including how common textbooks and style manuals for technical writing prioritize teaching correctness and conventional styles that conform to standard Englishes (e.g., Standard American English, Academic English, White Mainstream English) and "clarity" as a powerful metaphor for describing writing that works.

In addition, other common words and phrases in the above statements are evidence of new elements of the Communication Metaphor. For example, "short and to the point" and "brevity" are outcomes of an ideology of expediency. In short, expediency, the imperative to get more done more quickly with the least cost, has driven the worlds of industry, academia, and other technical contexts over at least the last two hundred or so years since the Industrial Revolution. Chapters 3 and 4 will develop how our common language about successful technical writing are supported by ideologies (tacit or professed beliefs about the world) and epistemologies (tacit or professed theory of what truth and knowledge is and how we know it) that have consequences for worldviews and languages that they exclude.

For example, the roots of *expediency* and its consequences will be developed in detail in chapter 4.

Looking systematically at the comments scientists and technical professionals make about writing and communication renders visible how conventional language about technical writing contains evidence for the Communication Metaphor. Systematic methodologies used by language and writing scholars for finding meaningful patterns in textual data are called "coding." Coding refers to the practice of assigning thematic categories to words or phrases in a chunk of text, such as a written survey response. In my analysis of written comments,[44] I assigned codes to the elements of the Communication Metaphor to create a coding scheme. For example, looking more closely at just one of the comments by "coding" it for elements of the Communication Metaphor reveals that it is informed by a complex, multivalent set of values about writing. In the case of Statement 1, the codes, or elements of the Communication metaphor, that have been assigned to the statement are style, expediency, and windowpane theory.

STATEMENT 1

Statement broken down by word or phrase	Element of the Communication Metaphor (code)
Short	Style
and to the point	Expediency
with facts only, no opinions.	Windowpane theory

In Statement 1 a "code" has been assigned to each word or phrase in the statement about what makes technical writing successful. What the codes mean is determined by the coding scheme, which explains when a code, or categories, should be assigned to a word or phrase. In the case of these three elements of the Communication Metaphor, the codes should be applied in instances of:

- **Windowpane Theory (WT):** When language, writing, communication, text, or documents are construed as unobstructed, neutral windows out to or into the real world of truth. The words *clarity* and *clear* are the most common markers. Also, *accuracy*.

- **Expediency (E):** When the segment mentions or invokes the value for writing and communication that prioritizes ends over means; ease, efficiency, or quickness of reading or comprehension for the reader/audience, or the outcome of writing/communication as the measure of its success.

- **Style (S):** When techniques for achieving characteristic professional and technical writing are mentioned or alluded to. For example, conciseness, succinctness, parallelism, plain language, organization, accessibility, elegance, level of detail, formal/informal, and level of technicality (jargon).

Looking at other statements about what makes writing and communication successful makes visible additional elements of the Communication Metaphor. Statement 2 adds a new code, "rhetorical framework," which means that a word or phrase in the statement invokes the commonly practiced approach to writing of addressing a defined audience that has particular needs or expectations:

STATEMENT 2

Statement broken down by word or phrase	Element of the Communication Metaphor (code)
Accurate	Windowpane theory
elegant,	Style
informative,	Rhetorical framework
engaging	Rhetorical framework

In the coding scheme, the code "rhetorical framework" is applied when:

- **Rhetorical Framework (RF):** When audience is mentioned or invoked, when persuasion or strategies for persuasion are mentioned or alluded to, when the persuasive purpose of a document or communication is mentioned or alluded to, and when conventions for content or sections of documents.

Finally, a look at a longer and more complex answer to the question "What are the characteristics of successful written communication in your workplace or Profession?" make visible two more elements of the Communication Metaphor: the conduit metaphor and correctness:

STATEMENT 3

Statement broken down by word or phrase	Element of the Communication Metaphor (code)
Clarity is the most important element.	Windowpane theory
You often only get one opportunity to present an idea,	Conduit metaphor
and in many cases, you never get to meet the person you are presenting to (when you submit grants, for example).	Rhetorical framework
It must be immediately obvious what you are writing about	Expediency
and why they should be interested.	Rhetorical framework
Telling a good story is very important, since we are asking donors and partners to invest in the work we do, which is based on an unmet need in the patient community.	Rhetorical framework
Conciseness matters a lot,	Style
since we work with very busy professionals, including clinicians.	Rhetorical framework
Editing and proofreading are also critical because, even if an idea is good, poor presentation or sloppiness can look unprofessional.	Correctness
Especially if someone is investing large amounts of money in your work, they want to know that they will be well represented in the programs and resources that the public sees.	Rhetorical framework

In the coding scheme, the codes *conduit metaphor* and *correctness* are applied when:

- **Conduit metaphor (CDM):** When the process of writing or communication is understood to be one of sending meaning (ideas, messages) across a distance.

Examples:

- It's hard to get that idea across to him.
- Your reasons came through to us.
- Next time you write, send better ideas. [13]

- **Correctness (C):** When the value of adhering language use to conventions and standards is mentioned or alluded to. Includes editing and proofreading. Also, mentions of error in writing.

Chapter 3 will say more about the conduit metaphor as an assumed model for how communication works. What is important here is the fact that Statement 3 is coded for *six* independent elements of the Communication Metaphor, which makes visible how common, conventional language about writing and communication is actually quite complex under the surface. While conventional language about technical writing and communication might roll easily off of the tongue—we don't really need to think very long to say that successful writing is "short and to the point"—that is not the same thing as saying that conventional language is superficial or lacks depth of meaning, or power. Whether it is the value of achieving correctness to standard Englishes or cultural agreements about what make for good technical writing style (conciseness), a generous focus on what the audience wants (rhetorical framework), the need for speed (expediency), or cognitive models for how communication works (the conduit metaphor), each element of the Communication Metaphor has a reason for being there—it has a story that reveals something about why as a complex of tacit assumptions and practices for writing and communication, as a conflagration of merged language fires, the Communication Metaphor has so much potential to do harm.

Understanding *communication* as a black-boxed term and a nominalization that has evolved historically and that is rooted in multiple values, models, and ideologies that have their own stories and motivations has made visible that there is something untoward afoot with how we approach thinking about language use in scientific and technical contexts. These multiple red flags now turn our attention to a deeper investigation of the complex of language practices that is the Communication Metaphor. Now that we can diagnose the problem of the Communication Metaphor, we can turn in chapter 3 to looking inside of the *communication* black box to what it means at the conceptual level. What are these cultural assumptions that we broadly share about how communication works? What are the shared, tacit conceptual models for communication that shape everything we do with scientific and technical language, including how we teach it and how we determine if communication is successful? How are these conceptual models reflected in the language that we use to talk about communication? Now that the Communication metaphor is visible, we can start to unpack what it means.

Chapter 3

What Does the Communication Metaphor Mean?

What do cells, containers and conduits have in common? They are metaphors, of course! Some readers might be wondering why I have waited all the way until chapter 3 to get into the topic of metaphor when I've been calling the subject of this book the Communication Metaphor since the very first page. I've already established that the Communication Metaphor is a complex of language practices, conventions, structures, and ideologies with historical, linguistic, and cultural origins that structure conventional language use in scientific and technical writing. In this chapter, I'll explicitly add metaphor as another element of the Communication Metaphor.

From an argumentative perspective, there is good reason for my authorial restraint in addressing metaphor. In brief, I didn't want to confuse the construct of the Communication Metaphor with a discussion of how metaphor functions as an element of it. Instead, I first established in chapter 1 that the Communication Metaphor is a problem and that nearly universally accepted conventions and practices of scientific and technical language have caused harm and continue to do so today. Next, chapter 2 made the Communication Metaphor visible. It raised the red flag on the word *communication* by calling it out as a black box, or a simplified front end for a largely mysterious back-end process that is hidden and messy and that operates on broadly held agreements that we assume, incorrectly, are understood universally. Chapter 2 also diagnosed a problem with the word *communication* because it is a *linguistic* black box, or nominalization. A nominalization is an abstract noun formed from an active verb that can displace a complex process (e.g., to communicate) from its origins and turn it into a widely accepted term (e.g., *communication*) that is accepted as meaningful in language use without definition. This chapter will open the black box and shed light on the tacit assumptions and practices that the nominalized word *communication* hides, in particular those related to metaphor.

Metaphor

Two types of metaphor contribute to meaning making about communication in scientific and technical writing: literary and cognitive metaphor. Both types of metaphor productively structure how meaning about communication is made. However, metaphor, which is often sold as a literary and linguistic superhero that dazzles with its power to weave and layer meaning for the good of literary fireworks or of explaining a difficult concept—or both—also perpetuates harm. In addition to the positive work that metaphor achieves by creating relationships between two often very different things to enhance meaning, these relationships can also become dominant or hegemonic, occlude alternative meanings, and become substitutes for thinking.

Both literary and cognitive metaphor wield the superpower to structure meaning by foregrounding one way of making meaning and foreclosing on other options. For example, in this chapter we'll open up how while it is perfectly natural to refer to the basic observed unit of biology as "the cell" (a literary metaphor), or to say "It's difficult to put my ideas into words" or "It's hard to get that idea across to them" (the container and conduit cognitive metaphors), these statements belie powerful assumptions that can limit what counts as knowledge and who can participate in knowledge making, and how.

Literary Metaphor—Creating Equations between Unrelated Ideas or Objects

Most people have at least a passing familiarity with the concept of metaphor, possibly from a school English course. Most people understand metaphor in the literary sense, that is, as an artful figure of speech that efficiently, and often beautifully, equates two unrelated ideas or objects. The equation creates new or enhanced meaning that is often rooted in our experience of everyday life. For example, here are a few fairly common metaphors:

- blanket of snow
- roller coaster of emotions
- heart of stone

Literary metaphor is also useful for grappling with extremely difficult, and abstract, concepts that fail adequate explanation by using only direct language—for example, love. Shakespeare, one of the masters of metaphor in English, turned frequently to metaphor to make sense of the most complex of human emotions. In these examples he equates love to a road or trail or assigns it agency, as if love were a sentient being:

> The course of true love never did run smooth (*A Midsummer Night's Dream*, act 1, scene 2)

> Love looks not with the eyes, but with the mind, And therefore is winged Cupid painted blind (*A Midsummer Night's Dream*, act 1, scene 1)

Some readers may also have learned about metaphor in a scientific or technical writing course. Just as Shakespeare found metaphor to be powerful for grappling with explaining one of the most complex concepts and emotions in human experience, literary metaphor can also be a powerful tool for explaining abstract and complex scientific and technical ideas to nonexpert audiences. For example, when trying to understand "the enormity and alien magnificence of something like a supermassive black hole, and its cosmic context," Caleb A. Sharf, an astrobiologist, found it compelling to write:

> Imagine diving into a collapsing soufflé of intergalactic gas.[1]

Linguistically substituting the name of a dish considered by some to be a delicacy due to the tremendous difficulty of getting it right (soufflés tend to collapse, rapidly, except under perfect baking and timing conditions) brings the reader efficiently, experientially, and beautifully into the concept of a black hole. Sharf's line brings the reader into a couple of fairly everyday experiences—the physical sensation of falling and the anxiety of the tenuousness of making a successful soufflé and the letdown for the bakers or eaters when they collapse. With the help of these equations between everyday experiences and the formation of a black hole, Sharf has been able to enhance the explanation for a general audience in a way that a straightforward scientific explanation might not:

> When [a dying star] has exhausted the internal thermonuclear fuels in its core at the end of its life, the core becomes unstable and gravitationally collapses inward upon itself, and the star's outer layers are blown away.[2]

Certainly, the encyclopedia explanation of how a black hole forms is at the level of general knowledge aimed at a nonspecialist audience; however, it still relies on the reader having some prior knowledge of several scientific terms: *internal thermonuclear fuels, the core, gravitationally collapses,* and *star's outer layers.* The encyclopedia text, while straightforward, still keeps the uninitiated reader at bay from comprehending the "the enormity and alien magnificence" of a massive star death ("star death" is, of course, metaphorical as well, since it implies that stars are "alive" like biological beings).

While the encyclopedia article relies on scientific jargon that may not be generally understood, the metaphorical explanation, however, also carries with it some risks. Readers who have no prior knowledge of the baked, intended-to-be-poofy whipped egg-based dish that is a soufflé, even indirectly, may not make any meaning from the metaphor. Experts might feel that the metaphor of the soufflé introduces untrue or misleading ideas about black holes that more exact scientific language would avoid doing. And both types of readers are right—using metaphorical language to explain technical topics is always risky because metaphors, by definition, foreground one aspect of a concept or object while obscuring others. For this reason, metaphors should always be used carefully and critically when explaining technical topics.

Explanatory metaphors used to explain science are also not always so easy to recognize as a "soufflé." For example, the word *cell*, a basic unit in biology, is a metaphor. This term came around because of Robert Hooke's publication of *Micrographia* (1665), a book about his observations through a primitive microscope. Hooke named the small compartments that make up plants *cells* because of their resemblance to the rooms that monks live in at a monastery.[3] It's fair to say that it's been a long time since users of the term *cells* have connected it consciously to the spare rooms occupied by monks, not least because few modern people have any experience of either monks or a medieval monastery. Yet the word persists in biology without its original metaphorical roots.

And the persistence of *cell* over the centuries matters because it has shaped how scientists conceptualize and practice biology. The cell is one of the foundations of a reductionist biology, which seeks understanding by taking things apart. This approach is driven by the assumption that how things work is a sum of their parts. While reductionist biology has revealed a lot about the biological world, it risks missing the forest for the trees.[4] In recent decades a new framework, systems biology, has emerged as an alternative way of thinking about the relationships between the parts of biological systems. Systems biology is big picture–oriented—it is about looking for connections and relationships rather than breaking things down into smaller and smaller elements.[5] With a new way of thinking about knowledge, or epistemology, biologists can research answers to different kinds of questions that the reductionist model was not able to answer.

Metaphors powerfully shape thinking, which is why they should always be used critically when called on to explain technical concepts. They can emphasize a wrong aspect of a scientific concept, but they can also die and take on zombie characteristics. Metaphors that we use often but no longer think about, such as the word *cell*, are called "dead metaphors." A dead metaphor has "petrified"[6] into a *cliché* or is no longer recognized as a special case of language use. But dead metaphors don't lose their power to shape our thinking about something. As we would be with zombies rising from the grave, we should always be wary of the lurking prevalence of metaphors in our language. Diedre McClosky was writing about how language about economics is heavily metaphorical (e.g., the invisible hand of a free market economy) when she stated: "Unexamined metaphor is a substitute for thinking—which is a recommendation to examine the metaphors, not to attempt the impossible by banishing them. . . . Metaphors evoke attitudes that are better kept in the open and under the control of reasoning."[7] Bringing unexamined attitudes and assumptions into the open is what this book is all about: may we all seek to identify our substitutes for thinking!

Conceptual Metaphor—What Our Language Says about How We Think

Now that we've established that metaphor is a commonly understood and practiced meaning-making tool in both literature and scientific writing, we're going to talk about metaphor in a slightly different sense that will make even more visible how our metaphorical language about communication foregrounds one highly conventional meaning about how it works while occluding other possibilities. To do this we are going to draw on a field of language study called "cognitive linguistics." Cognitive linguistics fundamentally argues for the hypothesis that language is governed by general cognitive principles,[8] including that both the grammatical structure of language and the words chosen to make meaning are governed—or controlled—by largely unconscious conceptual structures of the mind. More recently, neurological research has also begun to show connections between cognitive processes involving metaphor and specific parts of the brain[9]—that is, there is a mind-body connection, and it invisibly shapes how we think, talk, and write.

Metaphor, of course, is one of these conceptual structures of the mind. Like literary metaphor, conceptual metaphor works by mapping characteristics of one thing (the source domain) onto something usually quite unrelated (the target domain).

For example, a commonplace statement such as

"We have entered the twenty-first century"

is structured by the conceptual metaphor that

TIME (target) IS SPACE (source).[10]

In addition, a statement such as

"She demolished his argument"

is structured by the conceptual metaphor that

ARGUMENT (target) IS WAR (source).[11]

These examples make visible the general formula for a conceptual metaphor, which is,

TARGET DOMAIN IS SOURCE DOMAIN[12]

For our discussion of conceptual metaphor and communication, we are particularly interested in a conceptual metaphor called the "container metaphor." For example, these common statements are structured by the container metaphor:

He *fell into* a depression.

I'm slowly *getting into* shape.

They're *in* love.[13]

Which are structured according to the formula:

STATES (target) ARE CONTAINERS (source)

The container metaphor, as just one example of conceptual metaphor, is a very big idea and can be quite startling when encountered for the first time. Conceptual metaphor makes us aware that how we use language every day is governed at an unconscious level by assumptions that are deeply rooted in our experience of the world as beings with bodies. Bodies are containers with insides and outsides that are oriented in space and time and also by tacitly held cultural norms and mores. George Lakoff, a linguist who has been among the first and most influential in developing how cognitive metaphor shapes everyday language, reminds us that cognitive metaphor is not a matter of smart or pretty language choice, but that it has to do with how we understand the nature of being of things: "The metaphor is not just a matter of language but of thought and reason. The language is secondary. The mapping is primary. . . . The mapping is conventional; that is, it is a fixed part of our conceptual system."[14] What Lakoff is saying is that we can't escape conceptual metaphor when we use language—it is always operating at the unconscious level because it is so normalized into our language use. In addition, conceptual metaphor is not just a matter of the intellect but "governs"[15] our everyday actions, down to the most mundane details. The totalitarian ubiquity of conceptual metaphors speaks to both their stability and their utility for everyday language use but also to their power to privilege one conceptualization over another. If the states of depression, fitness, and love are fundamentally understood as containers in common language use, then they are also not conceptually structured as astral bodies, clouds, or horizons. If they were, we might commonly say or hear:

I'm slowly rising up to shape.

They are rising to love.

He rose to a depression.

Of course, it's not that we could never see these phrases—they are perfectly grammatical. But if we did, it is most likely that they would appear as a *literary* metaphor in a poem with purposefully unconventional language use or even as spoken language in a context in which these phrases would seem natural. For example, if you had parked your car in a parking garage with floors cutely named after positive emotional states (joy, love, happiness, ecstasy, etc.) then you might find yourself in an elevator with a friend "going up to love."

The Container Metaphor for Language

How we talk and write about language is also structured by conceptual metaphor. A prominent conceptual metaphor for language is the container metaphor. The container metaphor (fig. 3.1) extends beyond emotional (*in* love) or physical (*into* shape) states to include how we think and talk about language. The conceptual metaphor

LINGUISTIC EXPRESSIONS ARE CONTAINERS[16]

structures common statements about language:

> It's difficult to **put** my ideas **into** words.
>
> Try to pack more thought **into** fewer words.
>
> The meaning is right there **in** the words.
>
> When you **have** a good idea, try to capture it immediately **in** words.
>
> The introduction has a great deal of thought **content**.
>
> Your words seem **hollow**.
>
> The idea is **buried in** terribly dense paragraphs.[17]

In these statements, which could be pulled from any introductory technical writing textbook or handbook, linguistic elements, such as words, parts of documents (e.g., the introduction), or paragraphs are containers for objects, that is, ideas or meanings.

Figure 3.1. Words, sentences, paragraphs, document parts, and documents are containers for messages.

The Conduit Metaphor for Communication

While the container metaphor structures how we conceptualize the purpose of words, documents, texts, and other linguistic forms to *contain objects*, primarily thoughts and ideas, another conceptual metaphor structures how we understand how we move the containers of language around, or how we *share* ideas—in effect, how we send and receive messages between ourselves, other people, and the world at large. This is the conduit metaphor. A conduit, in a common sense, is defined as "a natural or artificial channel through which something (such as a fluid) is conveyed" or "a means of transmitting or distributing,"[18] such as illicit payments or information.

Remember that a conceptual metaphor is structured by the following formula:

TARGET DOMAIN IS SOURCE DOMAIN[19]

So that the conduit metaphor, which folds the container metaphor into the process for how language works, includes these three metaphorical concepts:

IDEAS (OR MEANINGS) ARE OBJECTS

LINGUISTIC EXPRESSIONS ARE CONTAINERS

COMMUNICATION IS SENDING

66 / Busting the Myth of the Communication Metaphor

Figure 3.2. Messages are sent across distances in packages. The conduit metaphor emphasizes the pathway, or the conduit, that a message travels between the sender and the receiver.

The conduit metaphor emphasizes the pathway, or the conduit, that a message travels between the sender and the receiver.

Figure 3.3. Once put in a container, a message must travel across a distance between sender and receiver for communication to be successful.

The conduit metaphor structures common expressions that we use for talking about language. For example:

It's hard to get that idea **across to him.**

Your reasons **came through** to us.[20]

You'll have to try to get your real attitudes **across to her** better.

It's very hard to **get that idea across** in a hostile atmosphere.

Your concepts **come across** beautifully.

Next time you write, **send** better ideas.[21]

In sum, the conduit metaphor, which incorporates the container metaphor, rests on these four tenets:

1. Language functions like a conduit, transferring thoughts bodily from one person to another.
2. In writing and speaking, people insert their thoughts and feelings in the words.
3. Words, which are packages, accomplish the transfer by containing the thoughts and feelings and conveying them to others.
4. In listening and reading, people extract the thoughts and feelings once again from the word and document packages.[22]

The thoughts and feelings extracted from the packages are exactly the same as and unchanged from what was originally put in the package.

It's not just passing apples

Before saying more about the conduit metaphor, I want to take a moment to absorb a really significant, but often overlooked, point about communication made by linguist Michael Reddy, who developed the notion of the conduit metaphor. In 1979, Reddy made the point that writing and communication often feel so hard because we expect them to be easy! "It should be clear that the overwhelming tendency of the system, as viewed by the conduit metaphor, will always be: **success without effort.**"[23] In other words, it is embedded in our concept of communication that it should be easy, yet *it isn't*. Our expectations for how easily communication should work are foiled nearly every time we open our mouths or put pen to paper. And Reddy's point shouldn't come as a surprise, as we see in the following conventional comments from writers in technical fields about how success in writing can be measured by the ease with which messages are shared:

> I can grasp the ideas easily without trying to second guess.
>
> —professor and researcher in biology at a research university

> When other researchers or students understand it, few mistakes, few clarifications required.
>
> —instructor and researcher in psychology at a college

> I can understand it without too much effort.
>
> —researcher in education data science at a university

> When a dense topic is understood easily with minimal clarification/miscommunication.
>
> —graduate student of biology at a research university

Reddy's point is that the expectation that communication, by its very nature, can be, and actually should be, seamless and easy, and not just for the "good" writers and speakers but for everyone and all the time, is baked into how we conceptualize how communication works via the conduit metaphor. The source of this ease is the conduit metaphor's confidence that communication happens because our thoughts and ideas are actually, meaning *literally*, transferred from our minds into those of others via the transport provided by all-purpose shipping containers, such as words, paragraphs, documents, genres, and other linguistic elements. Remember, the container metaphor construes thoughts and ideas as *actual* objects that go into these packages, then are sent along a conduit to be taken out of the packages and put into the minds of the recipients, unchanged.

Here's another way to understand Reddy's point: The conduit metaphor assumes that communication is a bit like passing an apple from me to you, but only if I was standing right next to you. The apple has a shape, weight, texture, and color that I can experience empirically via direct observation. When I pass the apple to you, you can also experience the shape, weight, texture, and color of the apple via direct observation. Because we can pass the actual apple, and not just the idea or representation of an apple, we can be fairly confident that communication, or the passage of the apple from me to you, has taken place.

However, as we all know already, this is not actually how communication works! Thoughts and ideas, unlike apples, are not objects with an immutable materiality we count on to substitute for the messy, unreliable, and often thoroughly frustrating process of interpretation. In the apple-passing scenario, an apple is a member of the shared world that we are already standing in together, not a product of our minds, which are, by nature, confoundingly separate worlds. We can't actually just pass the apple when we want to share thoughts, ideas, and information. Instead we have to describe our thoughts and ideas using symbolic means (e.g, words or pictures) to stand in for the actual apple and then hope and pray that the recipient will interpret the symbolic representation in the way that we intended. This is to say that the sequence of events is not apple-from-me to apple-for-you, but instead apple-from-me to representation-of-apple to maybe-apple-for-you (but possibly a peach or plum instead!).

To illustrate the fallacy of conceptualizing communication as simply passing real, juicy apples between friends, especially when writing, Reddy created a thought experiment (see fig. 3.4) that illustrates a world of communication in which the players are not assumed to share the same world literally or the same worldview metaphysically, or, as we would say more generally, they are not assumed to share the same context. In other words, Reddy's machine accounts for the material fact that our minds cannot travel with our messages.

Reddy proposes a mechanical machine[24] that can only pass messages from the world of one communicator (e.g., Ed) and spit them out into another world (e.g., Curt) with no ability to broadcast to the recipient of the message any information about the world, or the mind, the message came from. We can imagine that the machine might look roughly like a bank of post office boxes at a shipping store, except that the PO boxes are very slim

Figure 3.4. Alternative model for communication to the conduit metaphor.

and can contain only slips of paper. Since the machine can pass only representations of things and ideas, not objects (like apples), if Ed wants to share his favorite fruit with Curt, he has to first write a description of an apple and draw a picture of a Granny Smith. He then passes the representation of the apple through the machine to Curt. While Curt might receive the message without incident, there are no guarantees that Curt will recognize Ed's description and simple illustration as an apple, or even as a fruit, if there are no apples in Curt's world. Ed, of course, has no knowledge of whether Curt has apples in their world. So, what is there to do?

In order for the communication event to achieve the intended success of mutual understanding, Curt will have to pass back a representation of their take on Ed's apple representations through the machine. Then Ed will have to send back to Curt a refined description and illustration that accounts for Curt's interpretation. Oh, if only Ed and Curt could just see into each other's worlds (or minds)! This refining and accounting and passing back and forth goes on until, maybe, Ed and Curt converge on a common understanding as their descriptions and pictures converge. However, other than the emergence of directly observable similarities between Ed's and Curt's messages as they pass them back and forth, they can never know for certain whether they actually share exactly the same knowledge that a Honeycrisp apple from New Zealand is Ed's favorite fruit. All we can ever know for sure is the content of our own minds and that our own minds don't travel.

Being subject to such a cruel machine that strips all mind material from messages is how we experience communication to actually work, especially when writing. Although we remain eternally hopeful that we can just pass apples, we get daily feedback that the apples have, in fact, not been passed at all; for example, when meetings are missed, feelings hurt, appointments forgotten, instructions not helpful, technical explanations mystifying, and papers rejected. And what Reddy really wants us to understand by spending time understanding the distinction between passing apples and the message-passing machine is that our most common language about communication is structured by concepts that are *fundamentally at odds with our experience!*

So, it turns out that this *is* big news: The conduit metaphor, while it informs nearly all of our language about communication, *is actually wrong.* In fact, Reddy had some harsh words for the conduit metaphor and goes so far as to compare the English language to a complex computer system that we don't use properly because we are so obsessed about furniture instead:

> [The conduit metaphor model] of communication objectifies meaning in a misleading and dehumanizing fashion. . . . It influences us to talk and think about thoughts as if they had the same kind of external, intersubjective reality as lamps and tables. . . . It is as if we owned a very large, and very complex computer—but had been given the wrong instruction manual for it. We believe the wrong things about it, and teach our children the wrong things about it, and simply cannot get full or even moderate usage out of the system.[25]

When a totalistic, fundamental, tacitly held, and immutable belief turns out to be wrong, there are, of course, consequences. To return to the theme of harm that had been threading through this book since chapter 1, it should come as no surprise that if our mental model of how language and communication works doesn't match reality, then we are not only in for endless frustration and confusion but also complicit in maintaining and perpetuating fundamental assumptions about and values for language use that are not only wrong but that also cause harm.

If we believe and act as if the natural state of affairs is that communication is seamless and effortless, like passing apples, it also means that difficulty in communication is an unnatural state of affairs that requires an intervention to correct it, often with a heavy dose of moralizing about the virtues of getting it right. No wonder so many people, and people in technical fields in particular, consider themselves to be "bad" writers and communicators! What a relief it might be to hear that instead of being a bad writer and communicator you are actually right where anyone could be expected to be! If the natural state of affairs is that communication is an unreliable, messy, and difficult process of creating and interpreting symbolic representations with very little hope that it will work in the ways that we expect, when it does work we should be absolutely, fabulously, and hilariously amazed.

Another way to explore the profundity of how our expectations for how communication works are wildly mismatched with our experience is to imagine an alternative. What if our language about communication reflected Reddy's infuriating message-passing-machine-metaphor rather than the easy, breezy conduit metaphor? Returning to the statements above from science students and researchers about how they know that communication is successful, let's see how they might look different if it is not assumed that minds can travel, or be shared:

Original statement

I can grasp the ideas easily without trying to second guess.

Non–conduit metaphor statement

I grasp the ideas well enough to build a working instrument in my lab.

Original statement

When other researchers or students understand it, few mistakes, few clarifications required.

Non–conduit metaphor statement

When other researchers or students understand it, they are able to build working instruments to run their experiments.

Original statement

I can understand it without too much effort.

Non–conduit metaphor statement

I can build an instrument for my experiment that works.

Original statement

When a dense topic is understood easily with minimal clarification/miscommunication.

Non–conduit metaphor statement

When a dense topic enables me to build a working instrument in my lab.

In this series of revisions, which may seem unremarkable at first glance, the second statements are informed by the metaphor of the message-passing machine rather than the conduit metaphor. This is because all traces of the expectation that the successful receiving of the message has anything to do with creating an equivalency between the mind of the author and the mind of the recipient have been erased. Instead, success is measured in purely pragmatic and local terms of what works or doesn't work in the world of the recipient without any acknowledgement of how this success relates to what was in the mind or experience of the sender. We occasionally also see this outcome-focused pragmatism in some textbooks about the purpose of technical writing, such as this sentence that discusses documents without referring to them as containers and instead focuses on the practical outcomes for readers:

> Instructions and guidance documents enable people to accomplish a specific task.[26]

Fortunately, even the message-passing-machine metaphor doesn't consign us to a life of isolation where there is no hope of any mutual understanding with others, but it sure does make it harder to achieve than we expect. Since an unlimited number of messages can be passed through the machine, senders and recipients can receive messages, build their instruments, write new messages with iterations on the information, and send it to the receiver, who might then build a different instrument successfully, make more modifications to the message, and then send it back. Over time and many messages, the possibility exists that the sender and the recipient can converge on a very similar instrument, even a better instrument than originally planned, and come to a point of mutual understanding—success! However, if the process sounds like it has the potential to be exhausting, it does. But we already knew that about achieving successful communication.

We talk about lines, not conduits, of communication

In English, we rarely, if ever, use the word *conduit* when we talk about communication. It's generally a technical word reserved for specific contexts of use, such as electrical conduit, or tubing, that protects electrical wiring as it threads through a building. This is because the conduit metaphor is a conceptual metaphor, not a literary metaphor. While the conduit metaphor shapes our thinking, and therefore our language, about communication, it rarely appears as a word in our talk about language. For example, the statement:

> It's hard to get that idea **across to him**

is shaped by the conduit metaphor (ideas are objects that travel across through space in containers) but does not contain the word *conduit*. A parallel statement that includes the word *conduit*:

> It's hard to get that idea through the conduit to him

sounds much less natural unless a much narrower meaning of *conduit* was part of the context of the statement, such as a construction site where electrical wiring was being threaded through conduit piping. But in this more technical world such a statement would still be unlikely, given that it's unlikely that the item to be threaded, or transported, would be an idea—it would more likely be a certain gauge of wire or a certain pressure or volume of water or another fluid. In this case the word *conduit* is not functioning metaphorically because the word *conduit* refers directly to the object it defines rather than being a substitution for something else in order to add layers of meaning via comparison.

However, there is a word that functions as a literary metaphor for the cognitive conduit metaphor: *line*, as in *lines of communication*. *Line* is such a beautifully simple word that speaks right to the heart of the most compelling aspect of the conduit metaphor: the directness, and, as Reddy points out, the ease, of the unimpeded travel of ideas as objects. Mathematically, a line is a one-dimensional set of points that extends in opposite directions infinitely. A line segment is a portion of a line between defined endpoints, or locations, such as points A and B. Defining locations opens up the possibility of traveling between them and also the metaphorical application to transportation and communication. The line is one of the most basic elements in geometry, and, notably, fundamental to the art of drawing as well. This means that line is a fundamental ordering concept of our visual and conceptual worlds—it then seems natural that it would have great metaphorical power.

Outside the abstract world of geometry, a line can take the form of a myriad of things: a river, a railway, the mail, a ferry, a telegraph, a bus, a telephone, a rope or cable, even pigeons, as uttered by the chief of signals of the British Army: "If it became necessary immediately to discard every **line** and method of communications used on the front, except one, and it were left to me to select that one method, I should unhesitatingly choose the pigeons."[27] Notably, *line* enters our vocabulary when we talk both of communication and of transportation, a confluence of meaning that makes sense in the context of the historical evolution of communication as dependent on modes of transportation, as discussed in chapter 2. When lines of communication were literally roads and waterways, communication and transportation were both physically and metaphorically codependent. In this example from an account of the military efforts of the Civil War from 1895, modes of transportation and technologies of communication are treated as equivalent: "For two or three days we had no mail, no telegraphic messages, and no railway travel. Our only communication with the outer world was by steamer from Georgetown, D. C., to New York."[28]

In our daily lives, we refer to both *telephone lines* and *train or bus lines* without much thought, even though it's not clear what the word *line* refers to. On the one hand, in both of these cases *line* refers metaphorically to the physical manifestation of telecommunication or transportation systems: telephone cables can appear as an infinite set of points extending infinitely in opposite directions, or as a line, when they are hitched along poles. Trains actually run along parallel lines of steel, called "train tracks," that can appear to disappear into the horizon. But when we refer to *train lines*, for example, we don't only or necessarily mean to refer to the actual train tracks. The word *line* also refers to bus or train routes that appear on paper (or digital) route maps as lines, or, more accurately, as line segments with a route beginning point and route endpoint. So, a bus *line* is simultaneously a drivable route along city streets and a representation of those city streets drawn on a route map. But when we say *bus line* do we really distinguish between those two instantiations? Probably not, because we don't need to to be understood in most situations. *Line* is a metaphor that powerfully reflects our conceptual notion of how transportation and communication systems are ordered.

It speaks to the deeply intuitive resonances of *line* as a metaphor for the conduits of communication that it appears to predate the invention of the telecommunication technologies in the nineteenth century that most obviously beg the description: the telegraph (patented in 1837) and the

76 / Busting the Myth of the Communication Metaphor

Figure 3.5. The word *line* refers to the straight tracks of train routes and the representation of routes that appear on paper (or digital) route maps.

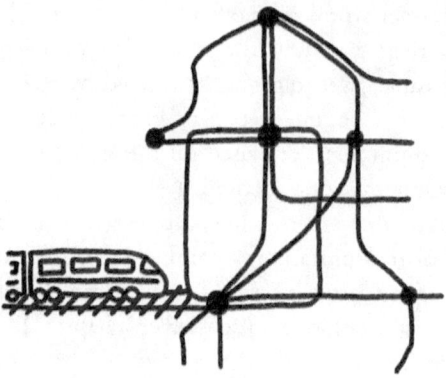

telephone (patented in 1876). A search of the Corpus of Historical American English turned up two cases that predate any widespread notion of electronic communication technologies. In the first case, from an account of the Revolutionary War, *line of communication* referred to the connection between two places. While it is unclear whether *line* refers to a continuous border or a road or other connecting byway, it definitely refers to a feature of the landscape: "As this fort lay on the great **line of communication** between Camden and Charleston, its fall was a great loss to the enemy."[29] In another example, *lines of communication* clearly refer to railways:

> Some of the other **lines of communication** promise similar results to the tracts of country which they are intended to serve. . . . The next line of rail-road from the Atlantic to the Western States, is the New York and Erie Rail-road.[30]

An interesting example from 1892 metaphorically links lines and communication well outside the realm of telecommunication technologies: this text equates the efforts of putting the mind of a deaf-mute child into communication with others to the lowering of fishing lines into the sea. This text is aspirational for the conduit metaphor: the child will be "drawn up into the light" into unimpeded communication with others:

> He says that the first efforts at her instruction were like **letting down lines** one after another into the bottom of the deep sea in which her silent soul lay, and waiting for the moment when she should seize hold of them and be drawn up into the light. In teaching we are continually doing this. We **let down our lines** and wait.[31]

I don't know when *line* entered into English as a literary metaphor for the cognitive conduit metaphor. It can be said, however, that since the early nineteenth century the appearance of *line* as a metaphor for both modes of transportation and communication has signaled that our notion of how communication works is fundamentally based on the conduit metaphor. Communication is a matter of transporting mind material in containers from point A to B in a straight, unimpeded line. Trying to imagine an alternative conceptualization of communication is nearly impossible, although chapter 6 will give it a shot by imaginatively exploring a scientific and technical world without the Communication Metaphor.

The Windowpane Theory of Language

Metaphor also shapes our beliefs about how knowledge is made and how it relates to language. These powerful beliefs underwrite how we teach and talk about how to successfully communicate scientific and technical knowledge. The primary epistemological assumption, or theory of how knowledge works, supporting the conduit metaphor is that meaning and ideas exist independently of language. This means that, essentially, the more that words, sentences, and other linguistic elements can get out of the way by being correct, unadorned, and without bias, the more the truth about the world out there can shine through (or be successfully communicated). This epistemology is sometimes referred to as the "windowpane theory of language" because it assumes that language "provides a view out onto the real world," even if that view may be "clear or obfuscated."[32]

Figure 3.6. Language is a windowpane that we look through out onto the real world.

If the idea that everyday language use is structured by a theory of knowledge seems too abstract to be connected to everyday language use, consider how one of the most common and uncontroversial statements about what makes for effective language explicitly draws on the windowpane metaphor: effective language is *clear* language. We see this in the comments made by the survey respondents when asked what makes writing successful:

> **Clarity** is the most important element.
>
> In priority order: **Clarity,** objectivity, brevity.
>
> "Good" written communication should be **clear,** relevant, and tone-appropriate (according to the audience).
>
> Successful written communication must be **clear** and unambiguous.
>
> Again, I would say that **clarity** is critical.[33]

And, *of course*, scientists and technical professionals would all agree that clarity is the mark of successful writing, because it is taught in nearly every introductory scientific and technical writing textbook. For example, a basic textbook, titled *Writing Science Right*, frames its advice for achieving good style around the windowpane theory in Chapter 2: Achieving a Readable Style Learn Techniques for Clear, Concise, Active Writing with sections on how to

> Write Squeaky Clean Prose.
>
> Use Active Voice for Clarity.[34]

There may be, in fact, no truism about language that is more generally agreed upon in North American technical and professional English than the fact that successful writing and communication is "clear" communication. In addition to the survey respondents above, just ask a group of introductory technical writing students on the first day of class what makes for good technical writing—clarity and objectivity will be mentioned many more times than once. And where did they learn this truism? In a high school English class? Possibly, but that wouldn't even be necessary, since the windowpane theory of language metaphorically already structures how we think about everyday writing and communication as well. It is an inherited rather than an adopted theory of language.

The problem with the windowpane theory of language is twofold: First, if we believe that unsuccessful communication is due to a muddy, fogged-up, or shattered windowpane, then fixing the problem is a matter of cleaning or fixing the window rather than revisiting the message. In addition, drawing on the container metaphor again, if we assume that the packaging of a thought or idea is separate from the idea itself, then asking a writer to repackage an original idea "more clearly" or "more correctly" shouldn't seem to interfere with the nature of the idea itself—except that it does.

Finally, we know that the windowpane theory of language is inaccurate, or does not reflect our actual experience of how language works. Returning to Reddy's infuriating communication machine that allows only representations of things, not the things themselves, to pass between people isolated from each other, we note that it is noticeably absent a window. How convenient it would be if the poor souls could simply peep through a porthole to see with their own eyes how the others live. How much easier it would be to bypass all of the tedious back-and-forth of messages, slowly refining and converging on what is meant and what works. How much easier would it be if we could just see into, or read, each other's minds? Like the conduit metaphor, the windowpane theory of language is misleading because it reinforces our belief that effective communication should be, and can be, effortless: That sharing ideas and knowledge is just a matter of getting the words, the sentences, the paragraphs out of the way.

Clarity Realized: Simplified Technical English

Some readers of this book may be nodding along impatiently ("Yes, yes, I know") with my critique of how metaphors shape meaning in scientific and technical communication. If I had to guess, I might say that these readers work in industries, such as defense and aerospace, where the elimination of ambiguity and interpretation from language is a matter of life and death. Due to the high stakes of misinterpretation in the design, manufacturing, maintenance, and operation of aircraft, the aerospace industry has developed a controlled language with very strict rules to absolutely minimize the possibility of misinterpretation, especially for workers who speak English as a second language. This set of writing rules and a dictionary that fixes a single definition for allowable words is called Simplified Technical English (STE). Simplified Technical English is maintained via Aerospace and Defense

Industries Association of Europe as ASD Simplified Technical English, Specification ASD-STE100.[35]

Simplified Technical English operates on the assumption that if you can control every aspect of language use, then a text can have only one meaning. For example, you can avoid situations where an instruction could have multiple meanings: "Replace the bolt in the cover" can be interpreted to mean at least two different things:

1. Put the bolt back in the cover.

2. Remove the bolt and put a new bolt in the cover.

However, a user of STE would know that the verb *replace* can mean only "to remove an item and to install a new or serviceable item of the same type," as in "Replace the relay."[36] So, in this case, only the second interpretation is accurate. The verb *put* is defined in the STE dictionary as "to cause something to move or to be in a specified position or condition,"[37] as in "Put the lever back to its initial position." Via these controlled definitions, the dictionary establishes that the verb *replace* indicates actions that comprise both *put* 'and *remove* so that there can be no ambiguity regarding what an engineer or technician is meant to do to accomplish a task.

In theory, STE is the ultimate realization of the windowpane theory of language wherein the containers (words) of ideas have become so perfectly transparent that the ideas that they contain shine through unambiguously. To achieve this, literary metaphor of any kind is precluded by the dictionary of fixed, and singular, definitions. It is of course no secret that the aim of STE is to fulfill the windowpane theory of language: The aim of STE is succinctly stated by a user of STE in a headline of an article posted to LinkedIn: "Increase the Clarity of Text by Using ASD Simplified Technical English (STE)."[38]

In practice, STE works for its intended purpose in the industries that have adopted it as a standard for technical documentation. However, you can't completely stuff the interpretation genie back in the bottle, no matter how many writing rules and fixed definitions there are. A LinkedIn user responding to the post above pointed out in a comment how writing STE is a form of translation and therefore carries with it all the same risks of lost or changed meaning, just like when language is translated from English into German, or any other language.

Whether the translation of documentation into STE is done by a person or computer software, translation errors are collateral damage to the process. For example, the LinkedIn user gives the example of STE compliance software that translated the word *present* (as in *now*) into the word *current*, which has a technical meaning for electricians that is not at all related to our position in time. Since the adverb *now* is noncompliant in STE, the correct replacement would have been "at this time."[39] Since computer software can read and apply a set of rules just like a human can, the problem in this case is not an error in the consistent application of the rules—this will happen and can be corrected. The error comes about because of the problem of moving words outside of their original context—in the process of translation the original word *present* was translated into a synonym, *current*, that in a technical context has a completely different meaning unrelated to the original. How was the software meant to know the context of the text, that is the larger system of meaning that any instance of language relies on for its correct comprehension? It probably couldn't, at least not very well, as evidenced by the error.

The point is that STE works when it does because of the extremely narrow context that it operates in. A controlled language can work only when no shared context among users is assumed but instead the users' context is wholly defined by the controlled language's specification. In essence, STE is a formal example of the outcome of Ed and Curt's exchange of communications about Ed's favorite apple via Reddy's communication machine. As they passed messages back and forth, Ed and Curt slowly and iteratively established a controlled language that built a shared understanding of their respective worlds—at least enough to share the idea of an apple.

Simplified Technical English is beguiling both because it is the realization of all of our language and communication fantasies (CLEAR, EFFORTLESS, UNAMBIGUOUS!) and because it works. Except, the fact that it works well for the specialized industries that employ it also exposes its profound limitations and the conceit that it is founded upon. In most communication situations, taking the time to establish a shared context via creating a shared controlled language, whether via Reddy's machine or an international professional body, such as the European Association of Aerospace Industries, is not reasonable, even in just slightly broader contexts than the aerospace industry. In this case we are consigned to accept that Reddy is right: *Communication is hard not because we do it poorly but because to make it easy is not practical in all but the most specialized, high-stakes situations.*

The Transmission Model: It's Hard to Resist a Model That Confirms and Amplifies Your Biases and Values

So far in this chapter the focus has been on how literary and conceptual metaphors, as elements of the Communication Metaphor, shape our language about writing and communication. The container, conduit, and windowpane conceptual metaphors are the inevitable inheritance of speakers and writers of most Englishes, and, of particular interest for this book, scientists and other technical professionals working in English. They operate mostly unconsciously via our conventional language choices to shape how we believe writing and communication works. On their own, without any kind of explicit explanation in textbooks or English classes, or acknowledged cultural significance, these metaphors powerfully shape our beliefs and actions about communication—for better and for worse. However, metaphors about language can be even more powerful and pervasive—and take on great cultural significance—when amplified by the official credence of a published model that gains broad cultural visibility and dissemination.

This brings us to another element of the Communication Metaphor: the transmission model of communication. The transmission model, which is often referred to as the "Shannon and Weaver" model, is a mathematical model, but its origins are in the tacit cognitive foundations of the container, conduit, and windowpane metaphors. The transmission model emerged post–World War II from the burgeoning fields of electronic communication and information theory, and then spread widely via the cultural wave of technological and scientific optimism and expansion of postwar America. But, wait, we are getting way ahead of the story!

It wasn't inevitable that a mathematical model for optimizing electronic signal transmission would become a powerful model for human communication. However, in 1948 Bell Telephone Labs engineer Claude Shannon published an article, "A Mathematical Theory of Communication," in *Bell System Technical Journal*. In the article Shannon argued for a set of twenty-three theorems describing the "limiting relationship between system capacity, system transmission speed, and signal degradation, or "noise."[40] It's possible that this article about electric signaling would have remained within the sphere of telecommunications and the engineering aspects of communication if it hadn't been republished in a 1949 book, subtly renamed *The Mathematical Theory of Communication*. The book included an interpretive article by Warren Weaver, also a mathematician who was interested in machine language translation, that explained for a more general audience how Shannon's theorems about electrical signaling could be applied to a more general notion of communication. Thus was born what came to be known, somewhat incorrectly, as the "Shannon and Weaver" model of communication.

The model is best known as Shannon's graphic interpretation of the mathematics for the five elements of a "general communication system":[41] an information source, a transmitter, the channel, the receiver, and the destination (fig. 3.7). One way to understand the diagram is as "a transmission model of communication where the mathematical value of a communicative message is affected by decoding, transmission, and encoding."[42] It has been often reproduced in technical writing textbooks as the transmission model of communication:

The original model (fig. 3.7) consists of five elements:

1. An *information source*, which produces a message.
2. A *transmitter*, which encodes the message into signals
3. A *channel*, to which signals are adapted for transmission.
4. A *receiver*, which "decodes" (reconstructs) the message from the signal.
5. A *destination*, where the message arrives.

Figure 3.7. An illustration of the transmission model of communication: Shannon's graphical interpretation of his mathematical model for a communication system.

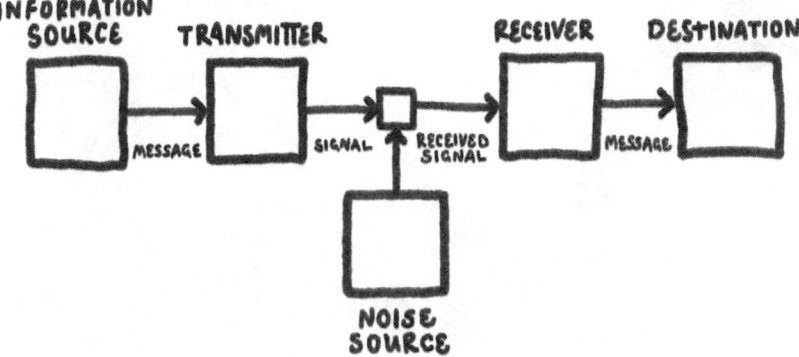

In addition, the model includes a sixth element, *noise*. Noise is the "dysfunctional factor,"[43] or one of the main sources of uncertainty or interference in the journey of the message along the channel to the receiver. A prototypical example of the transmission model would be a landline telephone system wherein the caller is the information source, the telephone handsets are transmitters and receivers, the electromagnetic wave is the signal, the telephone wire is the channel, and any noise would include electric or other interference to the signal or channel.

But, wait a minute, you might be thinking that the transmission model of communication looks a lot like the conduit metaphor, right? (see fig. 3.8). We might think that the transmission model of communication is simply a case of the conduit metaphor in the context of the technology of a telecommunications system, such as the telephone.

And, at a high level, an application of the conduit metaphor is exactly what the transmission model is and why it's not really surprising that it has been so influential in perpetuating the conception of communication as a process of putting information in containers so that it can be moved

Figure 3.8. Similarities to the transmission model confirm the deeply held biases of our tacit conceptual metaphors.

from one mind to another via a linking channel. That is, the transmission model formally confirms via technological progress the deeply held biases of our tacit conceptual metaphors, even though, as we know from Reddy, the conduit metaphor does not actually reflect real-world experience with communication.

Helpfully, the transmission model does go one step further than the conduit metaphor. Shannon and Weaver's model introduces the problem of noise or interruption to the message during the journey along the conduit. The transmission model does not assume that the default mode of communication is seamless and that containers passing along the conduit or channel are hermetically sealed. It does not assume that communication should happen without effort. In this sense the transmission model, which was born of the mathematical physics of information transfer, accounts for how when human communication is supported by technology the technology can introduce problems and distortions. In other words, the transmission model productively focused the problem of communication to be solved on the "noise." If solved, technologically noiseless systems would allow for more perfect communication.

Although in some ways it offers a more realistic model for communication, the main critique of the transmission model is that it reduces the problem of communication to a technological one with a technological solution, rather than accounting for how effective communication is actually a human problem of interpretation. It's a human rather than a technological problem, that the human experience depends on trying to understand what someone else, with their own mind, usually somewhere else, and with their own motives and situation, can possibly want you to understand and do. However, in a limited sense, Weaver understood this limitation of Shannon's model.

Even though Shannon's mathematical model was originally focused on the technical problem of how symbols of communication could be accurately and efficiently transmitted, Weaver argued in his 1949 article interpreting the transmission model that it was also relevant to two additional problems of communication: the semantic and the effectiveness problems.[44] In short, Weaver argued that how precisely the transmitted symbols convey the desired meanings to the receiver (the semantic problem) and how the received meaning affected outcomes or behavior in the desired way (the effectiveness problem) are also related to the technical problem of transmission. He went so far as to say, "The effectiveness problem is closely interrelated with the semantic problem, and overlaps it in a rather vague way; and there is in fact overlap between all the suggested categories of problems [including the technical problem]."[45] Weaver's intuition that a model of communication must account for both the technological transmission of data or information *and* the semantic transmission of meaning was right on. The complication of the relationship between the message, its transmission, and its reception is a helpful complication of the conduit metaphor, which does not account for the technological aspects of communication at all. However, the problems with Weaver's insights are at least twofold: (1) it isn't at all clear that Shannon's mathematics ever supported Weaver's generalization of the technical argument, and, at the very least, Weaver's interpretation depended on the reinterpretation of technical language in common terms that a self-respecting technically or mathematically trained person would never support, and (2) Weaver's insights, whether supported by Shannon's theory or not, have for the most part been interpreted in an oversimplified manner, and any subtlety that Weaver pointed to has been lost and was never represented in the transmission model diagram, which maintains many parallels with the conduit metaphor. In the fervor to generalize the model beyond electronic signal transmission, the controversies that made it interesting and potentially insightful were lost, or black-boxed, into a simplified interface for a complicated back end. In a haunting echo of chapter 2, this process sounds a lot like "black-boxing," ironically or not, a concept with its origins in cybernetics and information theory.

From a scholarly perspective, it's not clear that this erasure of the complexity of the transmission model matters very much. The fields of communication studies, writing studies, and other fields related to language research have justifiably and wisely long critiqued the limitations of the Shannon and Weaver model and moved on.[46] Interestingly, however, numerous alternative, more complex, models, such as the interaction and the transaction models that account for feedback, and psychological, social, and cultural contexts, have failed to take a strong hold outside of academic circles. If I ask introductory technical writing students to draw a model of how technical communication works on the first day of class, a lot of what I'll see will look like figure 3.7 or 3.8. Like it or not, there is something deeply aligned between the transmission model and our experience of communication in the modern world. Why is this? That is the interesting question here.

Rereading and processing Shannon's 1948 mathematical paper and Weaver's 1949 interpretation of it over seventy years later raises for me one question above all others: How and why did a theory of a *technological* process (Shannon's original aim) become a metaphor for the very *human* process of communication (Weaver's extension of Shannon)? What kind of a world is it that assumes that technological processes are metaphors for human processes? If anything, if the human condition has primacy in our concerns, shouldn't it be the other way around? What was it about America in 1949 that created the conditions for a mathematical model to take hold and to inform broad cultural mores about human communication? To ask this question is to ask about the ideological and rhetorical roots of the Communication Metaphor as a whole, which will be developed in chapter 4.

Conclusion

This chapter has unpacked the superpowers of literary and cognitive metaphor to structure meaning by foregrounding one way of making meaning and foreclosing on other options—all for better and for worse in scientific and technical communication. This is true for literary metaphors that emphasize one aspect of a black hole while overlooking others (a collapsing soufflé of intergalactic gas), that privilege one way of thinking about relationships within biological systems over others via dead metaphors (*cell*), and that reflect our deepest-held assumptions about how communication works (*lines* of communication). This is also true for the cognitive metaphors that structure our conventional language and thinking about writing and communication—the container, conduit, and windowpane metaphors tacitly preclude any alternative ideal of communication that might not be easy, unambiguous, or efficient but that might also be more inclusive of different types of language use and types of knowledge (epistemologies). Becoming aware of how metaphor both tacitly and purposefully shapes meaning, which has been the main aim of this chapter, begs the question of these alternatives. What might these alternative ideals be and how might they shape our everyday language use about communication if we were able to step outside of the influence of the Communication Metaphor all together?

This is a good place to remember that one of the central mysteries that motivates this book is how consistently scientists and engineers answer questions about what makes writing successful in their professions: 100 percent agree that writing and communication are important; words such as *clarity*, *conciseness*, and *objective* are nearly universal answers to what successful writing looks like. With the discussion of Standard Technical English in this chapter, it is tempting to suggest that such conventional talk about writing is its own kind of controlled language—and in some ways it is. Words such as *clarity* and *conciseness* have deeply shared, usually tacitly held meanings and reflect shared values within the context of science and industry. While, like STE, this shared language might work really well within the context that it was made to serve, its success also belies its major liability: that it excludes languages, metaphors, and models for communication that don't align with the rulebook.

Chapter 4

Where Does the Communication Metaphor Come From?

In chapters 1 through 3 I have argued that conventional statements about what makes technical communication successful, such as

> Good technical writing is clear, objective and can be understood quickly

are evidence of how the Communication Metaphor shapes our daily understanding and practice of writing in scientific and technical contexts. This statement is remarkable because of the powerful values and ideologies that it professes (clarity, objectivity, efficiency) but also because of how ubiquitous and normalized it is. Who wouldn't agree with this statement, even though, as I argued in chapter 3 in a discussion of the conduit metaphor, much of what it represents about how communication works is actually contrary to our experience of communication most of the time.

So, why does the Communication Metaphor work and persist even though it doesn't accurately reflect how language works for us? To answer this question, we'll need to ask about where the Communication Metaphor comes from. And this is a BIG question. I love to ask questions like this in my technical writing courses: "Where does the technical report format come from? Who thought it up, why, and when?" Questions like these usefully create some distance for students between the conventions of the technical report (the structure, the style, and the rhetorical aims) as tiresome rules to learn for a utilitarian purpose and the conventions of the technical report as outcomes of the evolving historical and rhetorical situation of science and technology over the past three centuries.

In this chapter we will explore the origins of some of the assumptions and theories about language and scientific knowledge that comprise the Communication Metaphor and also begin to consider who they serve. This chapter starts from the assumption that all conventions for scientific and technical writing conventions are an inheritance from the past that have been adopted and, sometimes, adapted, for the present. I also want to acknowledge that this inheritance, and its story, is a messy one and much longer than one chapter of a book. This chapter will not tell a comprehensive history, but it will tug a few of the narrative threads of the development of science and an industrialized economy, government, and society since the seventeenth century.

To initiate a discussion that is long, messy, complex, and, at times, controversial, I've organized it around two points of discussion that will bring into focus different aspects of the story. First, in continuation of chapter 3, we'll meet another metaphor for technical writing: technical writing as coin or currency for scientific knowledge. Second, we'll revisit one of the statements about what makes technical writing successful from chapter 2, that it is, "short and to the point, with facts only, no opinions." We will explore how and why this statement is shaped by one of the most powerful aspects of the Communication Metaphor: the ethic and ideology of expediency.

Another Metaphor: Technical Writing Is the Coin of the Economy of Scientific Knowledge

> Just as the most elevated science is *pure*, so, too, is the most accomplished technical writing *purified into neutral invisibility*. It is of these purified words that the clearest scientific knowledge is made. This purified scientific knowledge can then participate in an economy of knowledge and power—an economy unsullied by base words or ideas.[1]

These words, which summarize the main gist of a technical writing textbook from 1908, are chilling in their certainty—or at least should be. No four words better summarize the Communication Metaphor: writing "purified into neutral invisibility." Technical writing and communication as invisible has already been explored in chapter 2 (e.g., black-boxing and nominalization), but the word "purity" presents itself to us with fresh force. History has repeatedly borne out that no appeal to purity is ever fully benevolent or strictly utilitarian. Just invoking the notion of purity, a word that is loaded with moral, and arguably religious, mores, automatically creates the category of what is not pure, and, therefore, what must be excluded, stricken, or oppressed.

For T. A. Rickard, what must be done away with to make writing pure were slang, hybrid and foreign words, Latinate expressions, and literary meanings. As a mining engineer–turned-editor, Rickard forged his metaphor of technical writing as the currency, or coinage, of scientific knowledge and exchange[2] by equating impurities in language to impurities in valuable metal ores. In his view, a language rife with "impure alloys and cheap imports"[3] was a threat to science and engineering, and by extension, a threat to the human species. In other words, "Good technical writing ensures a genuinely sound currency for the scientific economy."[4] Within the metaphorical equation of purity and value (currency) with scientific progress and economic gain lie the roots of the potential for the conventions of technical language to exclude that which does not belong and to justify the exclusion on both pragmatic and moral grounds. In this metaphor of technical language as a valuable, purified metal, such as iron or gold, rings the origins of the logic that underlies much of the common sense about technical writing that we take for granted today: that it is "unflowery," "plain," "straightforward," "clear."

Rickard didn't invent the notion of technical writing as a currency that, in its pure state, expedites an economy of scientific knowledge for the benefit of society—even though he may have been among the first to write about it in an early technical writing textbook. Rickard inherited this notion from the seventeenth- and eighteenth-century thinkers of the European Scientific Revolution who powered the transition from the Renaissance to the early modern era. This transition was driven in part by Francis Bacon's (1561–1626) quest for the legitimation of scientific knowledge, empirical research methods, and rational thinking over the speculative, scholastic, and religiously inflected thought dominant until the sixteenth century. Bacon's struggle with tradition was followed by John Locke's (1632–1704) strong defense of modern empiricism. Together, they built the foundation for what would become the empirical methods, the scientific language conventions, and the state-sponsored institutionalization of modern experimental science in English.

In a nutshell, Bacon, followed by Locke, was deeply concerned about practical knowledge—that is, empirically derived scientific knowledge that can serve to improve the conditions of society. Bacon argued that knowledge generated via scholastic argument rather than the methodical observation of nature didn't generate useful outcomes or make life better for people in general. Instead, Bacon felt, scholastic knowledge rooted in the Aristotelian tradition of natural philosophy admires and applauds the "false powers of the mind"[5] without producing anything useful for the purpose of society or the increasingly bureaucratic and centralized English state. In contrast, the new science "aimed to connect progressive governance by the state to an ends-focused advancement of science"[6] by manipulating nature and improving the general living conditions. The logic goes that if science, and the communication of it, creates value for society, then scientific and technical writing is the currency that mints the value.

It is to the writings of John Locke however, that we look for an origin of the stylistic application of the value of purity in writing. For Locke, the language of the new natural philosophy of the recently founded Royal Society (1660) must be clear language that would allow scientific knowledge to flourish[7] and to create practical knowledge to benefit society. He specifically wrote in 1693 to theorize the function of language in establishing the authority of science and for communicating it: "Vague and insignificant forms of speech, and abuse of language, have so long passed for mysteries of science; and hard or misapplied words, with little or no meaning, have, by prescription, such a right to be mistaken for deep learning and height of speculation."[8]

If Locke sounds judgmental, to be fair, we can see where he is coming from in a text like this excerpt from *Arcana Microcosmi, or, The Hid Secrets of Man's Body Discovered in an Anatomical Duel between Aristotle and Galen Concerning the Parts Thereof* (1652):

> XIV. The *Aristotelians* make heat the efficient cause of the hearts publick motion: Others will have the soul; Others, the vegetive faculty; but *Aristotle* is in the right; for the soul works by its faculties, and these by heat; so that heat is the immediate cause of this motion, and the souls instrument; yet not such an instrument as worketh nothing but by the force of the principal agent; for the heat worketh by its own natural force, though it be directed and regulated by the soul; the heat then of the heart rarifying the blood into vapors, which require more room, dilate the heart; but by expelling some of these vapors into the arteries, and receiving also some cold air by the lungs, the heart is contracted, this is called *Systole*, the other *Diastole*: And as heat is the efficient cause, so it is also the end of this motion.[9]

We don't know if Locke had direct experience of this text, but scholastic argument (e.g., "Aristotle is in the right") as a scientific method, inflected by theological notions such as the soul's relationship to the body and complemented by individual observation of natural events, such as the beating of the heart, was still a strong part of the natural philosophy tradition at that time. We can see how Locke might have considered the science explained via this passage, "the height of speculation." It is worth noting, however, that the technical vocabulary about the contraction and dilation of the heart, *Systole* and *Diastole*, had been in circulation since the Alexandrian medicine of the third century BC and is still in use today—the roots of scientific language are many and complex.

When considering the origins of the Communication Metaphor as a whole, the key transition in this early modern period was the role that language was understood to play in the creation and dissemination of scientific knowledge. If argument, or disputation, is essentially a scientific method, as practiced by the Aristotelian tradition of natural philosophy, then language and the writing and study of texts is the main stuff of science:

> Language/argument/texts=scientific knowledge.

In the mixed writing methodology of argument, theology, and nonexperimental observation of nature apparent in Ross's *Arcana Microcosmi* there is a freedom of thought and expression that feels antiquated in light of today's narrower "just the facts," "objective," and "clear" approach to scientific writing. However, if science is to serve society and have utilitarian outcomes, then it had better be made relevant, comprehensible, and standardized for the stakeholders who will harness its value. If it is the outcome, and not the argument, that matters, then language, with its abstruse vocabulary, messy grammar, and uncontrolled inclination to incorporate all dimensions of human experience, including emotion and spirituality, had better get out of the way. The way to get language out of the way is to "purify it into neutral invisibility," as Rickard insists in the opening passage to this section.

Getting language out of the way became central to the rise of the new science in the mid-seventeenth century. In contrast to the *Arcana Microcosmi*, consider a text from 1665, which strictly reports on the observation of "fact patterns in nature."[10] In Robert Hooke's *Micographia*, the scholastic discussion of positions and ideas has been replaced with a methodical reporting of Hooke's activities as he observed a piece of cork under the lens of his invention, the compound microscope:

> I took a good clear piece of Cork, and with a Pen-knife sharpen'd as keen as a Razor, I cut a piece of it off, and thereby left the surface of it exceeding smooth, then examining it very diligently with a *Microscope*. . . . I told [*sic*] several lines of these pores,

and found that there were usually about threescore of these small Cells placed end-ways in the eighteenth part of an Inch in length, whence I concluded there must be neer eleven hundred of them, or somewhat more than a thousand in the length of an Inch, and therefore in a square Inch above a Million, or 1166400, and in a Cubick Inch, above twelve hundred Millions, or 1259712000, a thing almost incredible, did not our *Microscope* assure us of it by ocular demonstration.[11]

Robert Hooke's fastidious, direct writing style, which begins with a report of his experimental methods and then proceeds to report on his observations supported by the empirical proof presented via the new mode of the microscope image (fig. 4.1), more closely resembles the methodology of modern-day scientific reporting:

Reported direct observations of natural phenomena=scientific knowledge.

Figure 4.1. Hooke's text: "[The cork] had a very little solid substance, in comparison of the empty cavity that was contain'd between, as does more manifestly appear by the Figure A and B of the XI. Scheme, for the Interstitia, or walls (as I may so call them) or partitions of those pores were neer as thin in proportion to their pores, as those thin films of Wax in a Honey-comb (which enclose and constitute the sexangular celts) are to theirs."[12] *Source*: Huntington Library, San Marino, California, call number 487000:0930.

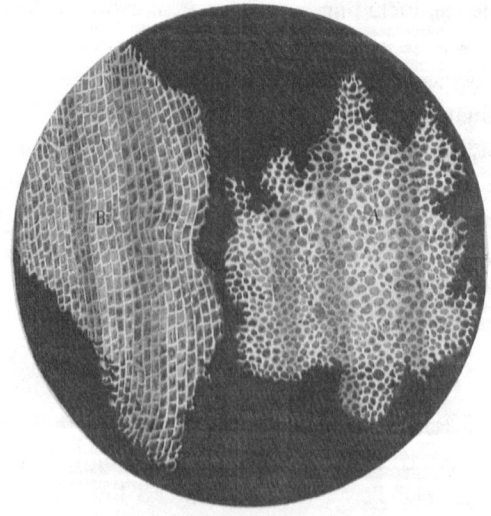

This is no coincidence: Hooke was an early editor of the publications of the recently founded Royal Society in England, which was founded to institutionalize and regularize experimental science and to realize Bacon's vision of experimental science as a project to benefit society and the state.[13] Hooke's relatively more concrete and descriptive style, as compared to the excerpt from *Arcana Microcosmi,* is moving closer to the writing style that avoids Locke's dreaded "Vague and insignificant forms of speech." Hooke's style is moving toward a "purity" in the currency, or language, of science by removing discussion that goes beyond observation of the phenomena or commentary about the method of observation. His writing feels more modern because the shift to limiting scientific argumentation to the reporting of methods and the results of direct observation via a direct, descriptive style of writing is a key element of our inheritance of the Communication Metaphor from this time.

Plain Style and the Royal Society

Hooke's writing reflects the style of science writing vigorously promoted by the Royal Society, the oldest continuously existing scientific academy in the world. For the Royal Society, the goal of Plain Style, according to Thomas Sprat, an early historian of the Royal Society (1667), was "a close, naked, natural way of speaking; positive expressions; clear sense; a native easiness; bringing all things as near the Mathematical plainness, as they can: and preferring the language of Artizans, Countrymen, and Merchants, before that, of Wits, or Scholars."[14] For scientific writers of the time there was a lot at stake in adopting the Royal Society's assertive and unadorned Plain Style—most of all recognition and publication in the society's journals. The tremendous force of the incentive to adopt the society's style of writing can be seen in the evolution of a text that was rewritten twice in order to gain entry and acceptance:

Version 1. "Which to us is utterly occult, and without the ken of our Intellects"

Version 3. "to which we are stranger"

Version 1. "those abstrusities, that lie more deep, and are of a more mysterious allon"

Version 3. "the Difficulties that lie more deep"

Version 1. "those principiate foundations of knowledge"

Version 3. "The Instruments of knowledge"

Version 1. "Plato credits this position with his suffrage; affirming"

Version 3. "Plato affirms"

Version 1. "is but the Birth of the laboring Mountains, Wind and Emptiness"

Version 3. "stands yet unresolved"

Version 1. "preponderate much greater magnitudes"

Version 3. "outweigh much heavier bodies."[15]

These excerpts show how the text was revised multiple times to achieve a new, plainer language style. This was achieved by removing metaphorical language, reducing the word count, using more concrete language, and adhering to a stricter subject-verb-object sentence structure.

The stylistic priorities of Plain Style may feel familiar to many readers, especially if you are in a technical writing course or have taken one in the past. This is because these stylistic priorities have been inherited today in what is now taught as "plain language writing" in technical writing textbooks. The society's direct, utilitarian Plain Style aligns with our contemporary notion of plain language writing in that it aims to expunge vague or unnecessarily complicated language from scientific and technical language in order to prioritize precision as the primary stylistic aim. However, while there are overlaps in the stylistic aims of seventeenth-century Plain Style and twenty-first-century plain language writing, their overall objects are very different: in the seventeenth century, the aim was to create legitimacy for science by forging a utilitarian style that differentiated the new empirical science from the old Aristotelian scholasticism that could serve an institutionalized science with a service to society; in the twenty-first-century plain language writing aims to make scientific and technical knowledge accessible to nontechnical audiences via the creation of texts, such as patient-facing medical information brochures or websites, that are largely separate from those produced by professional researchers. Critically, the stylistic priorities of the society's Plain Style over time *retained* the emerging specialized language of science, or jargon, which was then, and still is now, accessible primarily to trained scientists. In contrast, contemporary plain language writing is generally motivated by a desire to broaden the audience of scientific and technical writing *beyond* trained scientists by making the language more accessible. Contemporary plain language writing aims to reach more general, often vulnerable[16]—such as patient or immigrant—audiences by actively *reducing* the amount of specialized scientific language, or jargon, via techniques such as definition, illustration, and even welcoming back metaphor as a method of explanation for difficult concepts (see chapter 3). In practice, the aims of Plain Style and plain language writing are actually moving in opposite directions.

The Plain Style adopted and promoted by the Royal Society was a harbinger of Rickard's ideal of a technical writing "purified into neutral invisibility" and the harm that it would do and continues to do today. That this style would necessarily exclude nonconforming styles of thought and writing was not lost even on writers of the time, especially women scientific writers who were excluded from membership in the Royal Society due to their gender. In fact, feminist scholars argue that Plain Style was explicitly masculinized by the Royal Society as chaste—that is, pure—via the stylistic expulsion of any language or argument that appealed to emotion and experience, or the rhetorical appeal of pathos, that might interfere with reason.[17] In fact, women writers at the time, even though they also wrote their work in multiple styles, including at times the society's Plain Style when practical for their own purposes, recognized that Plain Style "censures"[18] the audience that responds to it.

The Royal Society's censure of the scientific community was deliberate, and unapologetic: writers and audiences who are not sufficiently rigorous or rational, or capable of handling "severe reason,"[19] were simply not admissible to this new sphere of standardized, institutionalized science. A major consequence of the Royal Society's assertion of exclusion was that it fundamentally changed the nature of the agreement between science writers and readers. As opposed to the burden of explanation and comprehensibility being borne by the writer, as might seem natural, the burden of comprehension shifted squarely to the reader alone. In short, as science became more institutionalized and professionalized, it was up to the reader to gain the scientific expertise necessary to comprehend a scientific text, not up to the writer to hold the hand of a less expert reader. The result was that only scientifically trained audiences could participate in science. This shift, however, did not go unnoticed or without critique. Margaret Cavendish, a scientist and polymath of the seventeenth century, showed explicit concern for the exclusivity of the Royal Society's Plain Style and argued that it should be inclusive of a broader array of readers.[20] She argued that if you are going to use jargon-heavy language that only scholars can understand, you might as well write in Latin. You might as well explicitly limit your audience to those with access to the education and position in society required to participate fully.

Cavendish wrote in favor of a scientific prose style inclusive of a broader audience, perhaps more akin to the contemporary notion of a plain language writing that aims to reach a broader, less expert audience. Cavendish's position was also in line with the aims of a scientific establishment formed upon Bacon's value of practical knowledge for the benefit of society. Cavendish said, "You must explain those hard words, and English them in the easiest manner you can."[21] Cavendish's viewpoint, however, did not win the day, and nonscientific audiences and the general public have been largely excluded from institutionalized scientific discourse[22] ever since. You can prove this to yourself by looking up a contemporary scientific research article in nearly any scientific journal—one outside of your field of specialty if you are a scientist.

The consequences of the Royal Society's regulation of scientific language have been significant. The willingness of the Royal Society to promote an exclusive style of writing that placed the burden of comprehension on a highly trained reader normalized centuries of self-censorship by generations of wanna-be scientists, concerned citizens, indigenous scholars, women, marginalized and minority populations, and all others who faced barriers to conforming, whether consciously or unconsciously, to the norms of a currency regulated by a seemingly genderless but still masculine rationality.[23] Within this masculine norm, commonly heard self-censoring comments regarding interest in and talent for science and technical fields, such as "my mind/brain just doesn't work that way," suddenly take on a new literality. These comments no longer reflect a problem with low self-confidence but an honest recognition of being excluded. The utterers of this comment are correct: the development of institutional science purposefully excluded types of knowledge making and writing practices that didn't fit and are still excluded today.[24]

Scientific Language Excludes: How Nominalization Enables an Increasingly Abstract, Cumulative Body of Scientific Knowledge

One way that science actively excluded nonspecialist audiences at the language level can be seen in the increasing use of the linguistic structure of nominalization—often recognized as scientific jargon—in scientific prose. Scholars who have studied the history of science writing have traced how the increasing use of nominalization in scientific language also enabled the evolution of science into an organized discipline (i.e., the Royal Society) that could develop new knowledge cumulatively and dialectically (via process of resolving controversies via debate). As we learned in chapter 2 during a discussion of the nominalization of the term *communication*, nominalization enables the transmutation of specific events and processes that happen in time into abstract concepts that can be generally referred to and transported across texts. While nominalization is a structure of all language use, it is also a process with special significance to scientific writing.

The excerpt[25] below from an astronomy text from 1822 traces how the process of nominalization can happen within a single paragraph via the development of the bolded words.

> The time of high water is principally regulated by the position of the moon, and in general, in the open sea, is from two to three hours after that body **has passed** the meridian, either above or below the horizon. But on the shores of the larger continents, and where the shallows and obstructions to the motion of the water, the interval between the time of the **moon's passage of the meridian,** and the time of high water, is very different at different places. The difference is so great, that at many places the time of high water seems to precede **the moon's passage.** For any given place, the time of high water is always nearly at the same distance from that of the **moon's passage over the meridian.**[26]

A closer look reveals the process of transmutation at the heart of nominalization. In the first sentences of the paragraph the scientific observation is encoded in a verb with supporting details in the first sentences:

> . . . the position of the moon, and in general, in the open sea, is from two to three hours after that body **has passed** the meridian, either above or below the horizon.

And then, once the background information has been established, abstracted, or nominalized, in later sentences:

> the moon's **passage** of the meridian
>
> the moon's **passage**
>
> the moon's **passage** over the meridian

Iria Bello argues that this type of nominalization of an event into an abstract concept, or jargon—the moon's passage—serves two cognitive functions. First, since the second part of the paragraph is built on the foundations of the first part, the text is able to build toward increasingly abstract ideas that can be repurposed later in the text without reiterating extensive detail or transferred outside of this text into another. This linguistic trick aids efficiency in scientific writing as single, abstract words or short phrases stand in for long, technical explanations. The second cognitive function for nominalization is to structure an overall progressive understanding of how scientific knowledge evolves: that is, new knowledge is built iteratively and mostly incrementally on existing knowledge via the transparent and deliberate steps of the scientific method.

The nominalization effect of abstraction and repurposing may not be obviously apparent in the passage about the moon, but across hundreds or thousands of scientific articles and one or two centuries, it's easy to see how necessary the process of nominalization becomes. Nominalization enables idea-driven, content-rich, efficiently written texts, which sounds like exactly what we need to report on new scientific research findings or to document complex technical systems. Linguists explain that nominalization has great utility in written texts, especially in scientific and technical writing:[27]

1. Nominalization is a characteristic of written language more than spoken language because written language is more concerned with abstractions and ideas.

2. Nominalization enables the organization of texts by ideas, reasons, and causes. Texts can be driven by ideas that exist outside of time rather than only by people doing things in time.

3. Nominalization adds lexical density (more words) to a written text, which enables written text to efficiently contain more meaning.

Together, these three functions for nominalization in scientific and technical writing reflect the historical development of an ad hoc, esoteric natural philosophy practiced by widely dispersed individuals into the organized discipline of science defined by a consensus around the legitimacy of an ever-expanding set of abstract ideas expressed using a specialized terminology, or scientific jargon.

Certainly, just as we can't reiterate our theory of how communication works every time we wish to refer to the concept in everyday life (see chapter 2), scientists can't re-explain every scientific concept in every text. However, the hazard of the compactness and efficiency that nominalization provides to scientific prose is the exclusion of readers who do not have access to the original background information and therefore may not be able to fully understand the scientific idea under discussion—this consequence of nominalization shows up when readers report that a text is so jargon-heavy that they can't follow it. It's as if the scientific enterprise is a process like that of building a brick house, where each brick, or nominalization, is a new scientific concept that has been accepted by the scientific community as new scientific knowledge. Once the brick is formed and added to the house, its origins become difficult to trace—where the clay came from, how the brick was formed, who formed it, and so on. Ultimately, the house, or even the castle, is so big that to the nonbuilder it has just always been there as a complete building—and may be more than a little bit intimidating to a visitor who doesn't know the family.

The brick castle analogy is apt given Bello's argument for the second cognitive function of nominalization in science. She argues that the linguistic process of nominalization that we see happening in the moon passage is a microcosm of how we cognitively understand the nature of scientific knowledge in general. That is to say, the nature of scientific knowledge is assumed to be largely progressive, in that scientific observation moves our understanding of natural phenomena from the specific to the generalizable, and new knowledge is built iteratively and mostly incrementally on existing knowledge via the transparent and deliberate steps of the scientific method. Bello argues that how we understand the nature of scientific knowledge is therefore encoded in the structure of how we write about it. In some sense, then, nominalization is the engine (or the the brick maker) that drives the advancement of scientific knowledge, and has done so in increasingly complex ways over the last four hundred or more years.[28]

Nominalization is an important element of what linguists call English Scientific Register,[29] which is also a component of the Communication Metaphor, that has been honed over time to serve the interests and purposes of scientific and technical communities. Nominalization is for technical and science writing, however, a double-edged sword. From the perspective of the study of English Scientific Register, a study of nominalization helps us to understand how scientific knowledge is formed and disseminated by writing. In addition, if we are willing to follow linguist Iria Bello down this path, the prevalence of nominalization in scientific writing is evidence of a powerful assumption, or ideology, about how scientific knowledge advances progressively, one observation turned concept serving as the foundation for the next observation turned concept. Without a doubt, the contributions of science to the development of the modern world have been extraordinary, which means that how scientists write about it has worked, and continues to work well, nominalizations and all. For our purposes here, however, Bello's argument about nominalization is an opening into a larger argument about how the structure of scientific and technical language reflects our deepest assumptions about what it is, how it works, and who gets to use it. Having the analytical tools to call out linguistic functions such as nominalization that might otherwise be invisible to us is the first step toward taking meaningful action for redressing the harms illustrated in chapter 1.

Admitting discomfort over the inaccessiblity of much writing about technical topics raises the question of whether it is in the nature of scientific and technical knowledge that they are generally off limits to anyone without specialized training. In other words, is it inherent to the nature of scientific and technical knowledge to also be linguistically inaccessible, thanks in large part to the increasing prevalence of nominalization in the scientific register? Or, is this turn of affairs the outcome of a history of powerful institutions, such as the Royal Society, creating scientific journals with powerful editorial boards that serve as the gatekeepers for what counts as knowledge and for who can make a contribution? But why did those gatekeepers develop and maintain conventions about language use in writing about science that conflated the credibility and authority of a text with inaccessible language—the outcome of the Royal Society's push for Plain Style? While it's definitely uncomfortable to mix the pursuit of scientific knowledge and the messy political barnyard of human affairs, we cannot ignore the fact that language is a tool of humans, who are by nature social beings, and thus our social priorities, biases, and personal interests necessarily shape how we enforce its proper use. In short, nominalization is not a neutral linguistic tool because it is used in service of human motivations that are shaped by how power is distributed in societies.

While nominalization may support the advancement of scientific knowledge, even while it also ensures increasingly jargon-heavy scientific writing, it can also have more directly insidious effects when used in technical writing that has real consequences for people in vulnerable situations. For example, consider this excerpt from a book aimed at informing pregnant women about what to expect:[30]

> Examination: You will be weighed so that your subsequent weight gain can be assessed. Your height will be measured. . . . A complete physical examination will then be carried out which will include checking your breasts, heart, lungs, blood pressure and pelvis.

Norman Fairclough, a prominent scholar of critical discourse analysis (CDA), a methodology of applied linguistics that aims to identify and reveal how power relationships are embedded in language use, discusses how the *Examination* text is an example of medico-scientific voice—a close cousin to English Scientific Register. Norman notes how a combination of nominalization ("weight gain," "complete physical examination") and passive construction ("can be assessed," "will then be carried out") positions the pregnant woman as an object subject to institutionalized procedures conducted by an anonymous agent. For Fairclough, the linguistic features of language, such as nominalization and passive verbs, are clues that point to the larger ethos, or culture and values, of the medical community. These language features add up to a text that allows for only certain "social identities," such as one where the patient is stripped of any of the life context that might have brought her to medicine and this text in the first place: an environment where the patient is a passive recipient of medical advice and expertise that belongs to an abstract and disembodied entity called "medicine" devoid of individuals with unique interests, biases, stories, or other characteristics of humanity.

The excerpt from the 1984 book about pregnancy is an example of a patient-focused text from medicine that shows us how language features such as nominalization and passive construction have the power to construct social reality. In this sense "social" reality doesn't only refer to the individual relationships that we form with the people important to our lives but also the institutional contexts within which we make those relationships. Institutional contexts, such as the medical profession, academic disciplines, the university, schools, and science—including the Royal Society—establish and maintain power relationships that are often based on inequality. We can see this when we consider the relationship between a doctor and patient, a professor and a student, a scientist and a graduate student, or a new employee and a manger. For a long time, up to and including 1984, and even today, the *Examination* text would likely be read as just a boring and dry but otherwise unproblematic, useful, and harmless informative text. Fairclough, and CDA as an area of linguistics in general, helps us to see how the language practices that appear to be simply pragmatic for accomplishing our daily business have broader implications for society, in particular how we grant some people more access and power to certain kinds of knowledge and practice and less to others.

The Consequences of the Dominance of Royal Society Plain Style for Women

In retrospect, it is no surprise at all that science became the province of the largely male and privileged sliver of the population that also set the original terms for whose language and style of knowledge creation and writing would be admitted. The censure of the audience for participation in science has had deep time consequences that are still abundantly evident today—350 years later. Women, and women of color, especially, were, both literally and metaphorically, written out of institutionalized, state-supported scientific and technical professions. Stepping into our time machine and zooming forward to the present, we can see these consequences in the demographics of employment in technical fields. While women are well represented in health-related jobs (74 percent—but includes nearly entirely female professions, such as dental hygienist) and the life sciences (48 percent), their participation drops off markedly in careers related to physical sciences (40 percent), computers (24 percent), and engineering (15 percent).[31] More specifically, only 20 percent of computer science and 22 percent of engineering undergraduate degrees in the US go to women,[32] although 58 percent of college graduates are women. In addition, only 18 percent of full-time employed software developers are women and only 9 percent of electrical and mechanical engineers.[33] Only 5.5 percent of employed engineers aged twenty-five to thirty-four years with bachelor's degrees are black women, compared to 9.2 percent for white women and 24.1 percent for white men, the most represented demographic.[34] Nowhere is the harm inflicted by the Communication Metaphor more empirically evident than in the persistent underemployment of women in technical and scientific professions.

At this point you may or may not agree with the argument that scientific and technical language style is complicit in the exclusivity of science and technical fields. Common objections to the facts of exclusion evidenced by employment statistics often rely on a brand of proud, even paternalistic, tokenism that points to the tremendous successes of a few superstar exceptions. I have experienced whiffs of this tokenism myself in casual conversations with engineering friends and acquaintances. In one case, a retired but still passionate systems engineer who worked for decades for one of the largest semiconductor manufacturing companies in the world shared with me his pride in having mentored several extremely talented female colleagues throughout his career. When I acknowledged his accomplishment but then pointed out that the real problem is the missing middle—where are all of the perfectly-competent-but-not-exceptional female engineers required to change the employment statistics?—he could only point to how his own two daughters, who both showed great interest in and an early proclivity for science and technology, eventually chose careers outside of science. Why? It's hard to say. Probably a combination of personal circumstance and preference shaped by an educational and professional milieux saturated with male-centric imagery, "masculine default" stereotypes, and unwelcoming culture.[35] Frankly, it's a problem, given the impossibility of reducing the problem to a single, fixable, cause, that has been purified into neutral invisibility.

Some readers might still consider it a stretch to draw a direct line between the scientific writing style normalized in the mid-seventeenth century and low rates of employment for women and other underrepresented groups in technical fields today. This is a fair objection, since correlation is never the same as causation, and the structural and systemic forces that drive exclusion are multiple and complex. However, as a scholar of writing, I also know and accept that language is powerful because it structures, reifies, maintains, and reproduces dominant assumptions, ideologies, practices, and ways of making and valuing knowledge (epistemologies). Conventions for the right way to write and speak all come from somewhere, and they always have an agenda, even if that agenda has been long forgotten. It actually *does* matter that scientific style was normalized back in the seventeenth century to exclude writers and audiences who couldn't keep up with "severe reason." While it would be too strong to argue that the patterns of exclusion in employment that we see today are direct descendents of the inherited conventions for scientific writing that I claim for the Communication Metaphor, they are evidence that the language problem is real and that exclusion will persist in spite of the best intentions for mentoring women by individuals such as my systems engineering friend.

Purified into Neutral Invisibility: On the Road to a Morally Justified Expedient Style

That the values espoused for science and scientific writing articulated by Bacon and Locke feel modern is evidence that the empiricists largely won the debate over the purpose and the methodology for institutional science, which is what we recognize as the university-, research institute-, and national laboratory-based science of today. However, as utilitarian as it was in Locke's time to harness language to serve the pragmatic ends of bettering society via science, and as much as it continues to be so today, we can't ignore the darker side of what it means to link the always morally inflected notion of "purity," or what we might refer to as "clarity" today, with what counts as a successful writing style. The language conventions that document and disseminate new knowledge will always be shaped to serve those ends and the stakeholders who benefit from them the most.

In Locke's, and ultimately Rickard's, appeal to purity lurks the ghosts and inheritances of the West's history of colonialism and the rise of English as the lingua franca (or currency) for technical and scientific pursuits. While the drive for a more standardized and stabilized scientific language positively served to shape science into practical knowledge that had utility for society, it also became the currency of a colonialist rationale founded upon the values of productivity, efficiency, and extraction. In addition to being one of the greatest philosophers in Europe at the end of the seventeenth century, Locke was also a colonialist who invested in slave trading companies and whose political philosophy contributed to the rationale for the appropriation of land from the indigenous peoples in North America by the British. Locke wrote, "God gave the world to men in common; but since he gave it them for their benefit and the greatest conveniences of life they were capable to draw from it, it cannot be supposed he meant it should always remain common and uncultivated. He gave it [land] to the use of the industrious and *rational*—and labour was to be his title to it."[36] Locke argued that because the British could extract more productivity from the land compared to the indigenous peoples who sustained an economy that allowed for natural resources to be replaced and maintained, British settlers were therefore entitled to the land.[37] Locke's philosophy equated labor and extractive productivity with the moral and legal right to appropriate land. He yoked English productivity to rational thinking as defined by the emerging institutionalized scientific establishment. In Locke's view, rational, scientific language must be free of "vague and insignificant forms of speech," in order that science can flourish and serve society, which explicitly included the vast economic colonialist project in North America.

It is hard to overemphasize what a brilliant, and also pernicious, move it is to yoke rational thought and scientific and technical language to the morally justified aims of productivity and efficiency. If economic productivity is established as a moral good with scientific, rational thinking as the way to that end, then scientific and technical writing is the currency of that productivity. By logical extension, both metaphorically and pragmatically, the purer and clearer language is, the better a handmaiden language will be to economic productivity. In fact, while the language of natural philosophy once had a kind of self-conscious performative element to it due to its more ornate style and synthetic reasoning, now the efficacy of scientific and technical language is explicitly linked to its invisibility, or "clarity." And if the seventeenth century was roughly the beginning of the power of the currency of scientific language in English and its complicity in epochal endeavors that did great harm, including genocide, it was far from the end. In Locke's thoughts about proper scientific language we see one of the philosophical origins of what rhetorician Steven Katz called the "Ethic of Expediency."[38] The expediency, or focus on outcomes over means, of technical language, he argues, enabled the Nazi regime's rationalization of the genocide of millions of Jews and other marginalized populations during the Holocaust.

A Return to a Key Element of the Communication Metaphor: Expediency

When scientists and technical professionals today point to "clarity, objectivity and ease of understanding" as the hallmarks of successful technical writing, they are citing their historical inheritance from the seventeenth century. This inheritance, broadly speaking, is the Communication Metaphor. Tracing the historical development of science and the Royal Society as an institutionalized profession with regulated norms for communication, what linguists now call English Scientific Register,[39] reveals how the socio-historic story of science is reflected in the structure of language scientists use to write about it today.[40]

Overall, the inheritance of English Scientific Register is a good thing for individual students and professional scientists and engineers. It would be impossible to maintain coherence and efficiency in scientific discovery if every researcher had to reinvent how to communicate about their findings or if competing or collaborating researchers adhered to different sets of rules, making the comparison or meta-analysis of scientific findings impossible. Writing studies scholar Charles Bazerman eloquently reminds us that scientific writers today stand on the shoulders of past innovators:

> If each individual writer does not think originally and creatively about how to master recalcitrant language in order to create such powerful stories, it is only because the genre [the scientific article] already embodies the linguistic achievement of the three hundred years since the invention of the scientific journal necessitated the invention of the scientific article.[41]

During the early days (1665 until around 1800) of the Royal Society's journal, *Philosophic Transactions of the Royal Society of London*, the style of document that we now know as the scientific journal article developed in response to the increasingly centralized and experimentally focused scientific community. During these 150 years, reports in the journal evolved from being recollections of observed events or cookbook-type recipes for reproducing scientific effects to what we recognize today as credible and substantive contributions to scientific knowledge. Due to this inheritance, science students writing today learn that the multiple sections of a peer-reviewed scientific journal article must credibly achieve well-defined purposes: contextualize an experiment within the research and controversies from which it proceeds, propose a hypothesis to be tested via a documented scientific method, discuss how experimental results contribute to generalizable knowledge, and respond to possible objections to or criticisms of the methodology or conclusions.

Thanks to the institutional and cultural pressure applied by the Royal Society, the evolution of the journal article as a publicly circulated genre of writing enabled scientific observations and discoveries to be shared, scrutinized, and reproduced by other scientists who were not physically present to observe the original experiment. Bazerman's study of the origins of the research journal article reminds us that specialized types of writing evolve, by design, over time to meet the needs of the community that uses it—even as, by necessity, specialized types of writing also exclude readers and writers who are not part of the community. But that's the rub: the exclusion of nonspecialist readers and writers from scientific writing is also an inheritance of the seventeenth century. Recognizing the conventional format of the scientific journal article as an inheritance is an opening into a larger argument about the origins of the structure of scientific and technical language and how it reflects our deepest assumptions about what it is, how it works, and who gets to use it.

While the first half of this chapter argued for how what we accept today as English Scientific Register, or the grammatical and linguist and style conventions for scientific and technical writing, have their origins in a battle over why to pursue science and the proper way to do and communicate it, the second half of the chapter will examine why the victorious Plain Style of the Royal Society took hold and persists today. This examination starts with remembering that embedded in statements about writing by contemporary scientists and technical professionals are values and assumptions with origins in the seventeenth century, and earlier.

The Royal Society would certainly approve of how a scientist today described what makes scientific and technical writing successful, for example, that it is "short and to the point, with facts only, no opinions."[42] This statement might sound familiar, as it figured at the end of chapter 2 as a statement that was coded to reveal the several elements of the Communication Metaphor that shape it:

STATEMENT 1

Statement broken down by word or phrase	Element of the Communication Metaphor (code)
Short	Style
and to the point	Expediency
with facts only, no opinions.	Windowpane theory

Of particular interest for this chapter is the phrase "and to the point" (row 2), which has been coded for the element of the Communication Metaphor called *expediency*. As explained at the end of chapter 2, the codes were defined to make visible how contemporary values for successful technical and scientific writing shed light on the origins of the Communication Metaphor. The code *expediency* is defined as:

> **Expediency (E):** When the segment mentions or invokes the value for writing and communication that prioritizes *ends over means*; ease, efficiency, or quickness of reading or comprehension for the reader/audience, or *the outcome of writing/communication as the measure of its success.*

In the description of the code for expediency, we can see the traces of Bacon's quest to bring science into the service of society (that it be outcome-based) and Locke's focus on unadorned reason (ease, efficiency, or quickness of reading or comprehension for the specialized reader/audience). If science, and the communication of it, is a means to the end of a better, more rationally based, society, then the language conventions that document and disseminate new knowledge will be shaped to serve these ends. In other words, expediency, as the logical justification, or ideological warrant (i.e., widely agreed on supporting reason), for the society's *Plain* Style of writing, yokes rational thought and the style of scientific and technical language to morally justified aims of productivity and efficiency. This yoking has had some unsettling consequences.

The prioritization of the ends of communication over the means of communication via a language style that enables ease and efficiency of reading may not seem particularly pernicious at first glance. In fact, it seems pretty reader friendly—who wants to waste time wading through extra prose that doesn't accelerate the reader to the main takeaway of the piece of writing? But this is exactly the point—expediency, or efficiency in service of productivity or other ends, is a central value of education, government, science, and industry in today's world. It's the currency that we—or those who have overall benefited from its ubiquity—use. Expediency is a value that has shaped the culture and the history of the Western world since at least the seventeenth century, and it continues to do so today. In fact, expediency is an ethical end in itself because of how it has moral weight: progress becomes a virtue in a technocratic, capitalistic culture, motivated by the "technological imperative."[43] It shouldn't surprise us that how we use language and talk about language reflects traces of expediency because language practices, or what linguists call discourses, structure both our sense of reality and our notion of our own identity.[44] It is difficult, and uncomfortable, to consider that what we assume to be successful writing and our membership in scientific and technical communities are directly connected.

In fact, technical language and expediency are intimately entangled. Untangling them will shed light on the origins of expediency as a core value, or ideology, for what is considered successful writing.

EXPEDIENCY AS A WEAPON IN THE HANDS OF MADMEN

The Roman rhetorician Cicero warned us that the powers of language and rhetoric in the hands of "madmen"[45] would have catastrophic consequences. While it might be tempting to turn to the numerous examples of hateful or divisive rhetoric from contemporary politicians or activists to show how this has turned out to be absolutely true, fiery political rhetoric is not the topic of this book; so-called staid scientific and technical writing is. However, in the hands of madmen, technical writing can also serve violent, evil ends, comfortably sheltered behind its mantle of expediency.

For example, at first glance, the following excerpt from a technical memo recommending changes to a van design successfully adheres to stylistic conventions for technical writing that achieve the ideal of "short and to the point, with facts only, no opinions": "For easy cleaning of the vehicle, there must be a sealed drain in the middle of the floor. The drainage hole's cover, eight to twelve inches in diameter, would be equipped with a slanting trap, so that fluid liquids can drain off during the operation. During cleaning the drain can be used to evacuate large pieces of dirt."[46] This example of technical writing is detailed, objective, and to the point. It efficiently identifies the recommended change to the vehicle design and gives the justification for it. Its style of language and argumentation strategy are functionally serving the technical ends of the document, which is to improve a vehicle design. A manager reading this document would not worry that their time was being wasted.

This is a comfortable line of thinking until we learn that this technical memo was written in Germany in 1942 within the context of the Nazi program to exterminate Jews and other persecuted groups. At this time gassing vans were being used to execute women and children because it was more efficient and "humane."[47] Improvements to the vans were intended to improve their efficacy as killing machines. Suddenly, we have a completely different understanding of words like "fluids" and "pieces of dirt." This is a shocking and upsetting turn in the analysis of this passage—so much so that it might actually turn your stomach; I know it turns mine. The discomfort is emotional and visceral—perhaps not a very common reaction to scientific and technical writing.

This discomfort requires a moment of pause here, first to honor the dignity and memory of the millions that suffered from the Holocaust, and then to justify my resort to shock value to make a point. I include this passage because it gives real meaning to Cicero's warning about placing the powers of language and rhetoric in the hands of madmen, which we are now forced to confront applies to technical writing as well. Shock value, I have learned, is required when teaching about the ethics of technical and scientific writing. This is because, in general, the professional ethics taught in standard technical writing textbooks does not account for how to limit the power of rhetoric in the hands of madmen, or, even more importantly, in the hands of anyone and everyone. This is because it's not just the programs, governments, companies, and causes that we choose to work for that determine the goodness of our actions, it's also the tools that we are presented with to use—tools, like professional and technical rhetoric, that we believe are by themselves neutral, when not sullied by an evil motive.[48] What I find truly troubling among students, and, frankly, among people in general, is the comfortable assumption that if they ever were complicit in evil, it would be so self-evident that they would, for sure, take a stand or leave their jobs. On the contrary, most evil happens via the everyday, banal activities of regular people who often have only partial knowledge of the impacts of their actions—and technical writing is right at hand as that helpful, ostensibly benign tool.

While it is tempting to restore our comfort by situating the source of the evil evident in this memo wholly in the unique context of the Nazi regime, rhetorician Stephen Katz, who brought this memo to the attention of scholars in 1992 in his journal article "The Ethic of Expediency," goes to great lengths to make a much more subtle argument that complicates our swift return to comfort. It's not that simple, Katz argues. He makes a much broader argument that shows how what he calls an "ethic of expediency" underlies *all* technical and scientific writing—not just Nazi writing—and that this ethic, which is so predominant in industrialized Western culture, was at least partially responsible for the Holocaust.[49]

That technical writing might be complicit in the Holocaust is an upsetting idea, and not all scholars have welcomed it.[50] But let's try to understand Katz's argument more carefully to take full stock of its implications. The argument begins with Aristotle's theory from the fourth century BCE in his treatise *Rhetoric* about the division of speech types into three categories: (1) judicial (the discussion of events that happened in the past), (2) epideictic (speech of praise and blame—such as funeral eulogies or, more broadly understood today, advertising), and (3) deliberative or judicial (the discussion of future events for the purpose of deciding on their merit or worth). Katz places scientific and technical writing squarely in the category of deliberative rhetoric, which makes sense in terms of how the purpose of technical memos, and other genres of technical writing, is to contribute to decision making, consensus building, or other actions that impact the future outcomes of events—and their effects on people. In addition, scientific journal articles or grant proposals are arguments for new scientific knowledge or for support of a new line of research—both of which impact the future actions of the scientific community.

If, according to Katz's reading of Aristotle, scientific and technical writing is deliberative rhetoric, does this also mean that deliberative rhetoric, at least the subcategory related to matters of science and technology, must also have a singular style? In other words, must all technical rhetoric be "short and to the point, with facts only, no opinions"? As a case in point, we can ask about the Nazi memo if its style is at all related to, or even determining of, its content recommending a redesign of the vans? Or, put differently, if the van memo were written in a different style or genre would the outcomes of the redesign have been different for people—possibly more humane? Is the message about the van redesign dependent on the style of the memo?

An initial, and commonsense, answer is no. This is the easy answer if we readily accept that the objective style of technical and scientific writing does not at all shape its content. Objectivity is the value that informs the windowpane theory of language (see statement 1, row 3), wherein the object under study is removed from the emotional-personal-political matrix of the author and the community and wherein documents serve as unobstructed, neutral windows out to or into the real world of truth. Objectivity demands that language is but the neutral, transparent messenger of technical truths—language must first of all get out of the way. In this value we also hear a strong echo of Rickard's admonition from the opening of this chapter that successful technical writing is "purified into neutral invisibility." Given a purified view of technical style, we would expect a memo detailing changes to a Nazi killing machine to look quite similar, stylistically and argumentatively speaking, to a memo detailing changes to a marshmallow Peeps–extruding machine at the candy factory. And given a purified view of technical style, this is not at all a problem; at least, it is not at all a language problem. While I find this fact quite discomforting, in my experience, most technical and science writing students are willing to accept that the van memo and the Peeps memo could exist in a similar form and style even though the impacts that they have on people are so catastrophically different. If we accept that there is but one style for successful technical and scientific writing, then we also readily accept that the context that determines the content of the document is independent of the style of the document—a technical memo is a technical memo.

And what is wrong with this argument? It is widely accepted as a default understanding of technical style—so the cultural foundations of it must be robust. What is wrong with it is that those cultural foundations are the ethic of expediency. In the case of the Nazi memo, technical style's commitment to objectivity enables a writing style that is void of language that humanizes the consequences of changes to the design of a piece of machinery. This is justifiable, at least in theory, because the desired outcome of the Nazi memo is approval to improve the design of a machine, and not an endorsement, at least explicitly, of the violence that the machine is capable of. This is expediency in its most extreme illustration. It is an argument in which the ethical necessity of improving technology (i.e., progress is virtue) is privileged over the means the technology enables. In addition, the credibility of the memo rests on the fact that its argument is technically correct because technical correctness is a moral end in itself—a value that is not limited to the individual who authored the memo but that is a general cultural and ethical norm in capitalistic culture.[51] Katz puts it like this: "When the ethos of deliberative rhetoric (ethic of expediency) is combined with science and technology it can create a moral basis that in the extreme enables a regime like the Nazis."[52] How handy that the memo's author, Just, had learned the conventions for writing a technical memo, possibly in an engineering training course years before. He had available a tool perfectly suited to his rhetorical task.

What we have learned is that the context of Hitler's regime is not the only source of the evil that motivates the Nazi memo; another source is the ethic of expediency, or the values and practices for technical writing that precede and succeed this period of history. Katz pushes us to go a bit deeper in unpacking the relationship of context, style, and ethics in deliberative rhetoric. In effect, he argues that technical writing and rhetoric, regardless of the context in which it is practiced, is ethically compromised. Katz can make this argument because Aristotle, who wrote down the first systematic general theory of rhetoric, understood language use and argument as a practical rather than a theoretical science; that is, rhetoric is embedded in and formed by the messy, contentious contingencies of human action rather than dictated by the rational rules of logic. This means that rhetoric is a *praxis*, or social action, that is shaped by *phronesis,* or the ability of the speaker or writer of rhetoric to reason about what is good for the community. What is good for the community is defined collectively. In other words, rhetoric, unlike logic, is an activity that centers and prioritizes human interests over abstract consistency. This would seem to make rhetoric, including deliberative rhetoric, well suited for preventing decision making that would cause harm to the community.

Except, as might be expected, given the very existence of the Nazi memo, there's a catch: If what is good for the community is determined to be utility or achieving ends over means, as we saw first developed in the writings about science of Bacon and Locke, then expediency becomes the ethical end of technical and scientific writing—not human concerns. The conventional acceptance of objectivity, efficiency, and productivity as outcomes of successful technical and scientific writing reflect the broad and deep commitment of Western culture to expediency as it has been shaped and cultivated by its predominantly capitalist and technocratic systems.[53]

At this point in the discussion, it might seem natural to turn to a critique of capitalism, the economic system that supports most technical organizations in the Western world, since, to be honest, a critique of capitalism already fundamentally underwrites Katz's ethic of expediency. However, I want to tread lightly in this part of the argument about the Communication Metaphor. Since I expect that many readers of this book have a present or future stake in industry, government or publicly or privately funded research that comprise a large part of the economic machinery of the Western technocratic-capitalistic system, I want to make clear that in my view it is not necessary to argue wholly against capitalism to adopt a critical view of the Communication Metaphor; or, as I ultimately aim to do, to propose an alternative to it. However, it is also not a coincidence that the history of scientific and technical writing, as narrated so far in this chapter, has largely happened concurrently with the Industrial Revolution in England (1760–1840) and the Western world. Critiques of the consequences of the Industrial Revolution offer some useful concepts for understanding the roots of the Communication Metaphor, and in particular the ethic of expediency.

Severe Reason Goes Military-Industrial: Technological Rationality and Hyperpragmatism

The philosopher, economist, and social theorist Karl Marx wrote his major works critiquing capitalism during an era (1840s–1860s) when advances in science and technology had promised to transform the world with new inventions, such as the steam engine and the full mechanization of the textile industry. His critique of capitalism made visible for the *first time* the central role for science and technology in human history[54] and how the new industrial capitalism had fundamentally reordered the world socially, politically, and economically. Whether or not an individual in the twenty-first century personally agrees with a critique of contemporary capitalism is less important here than the acknowledgement that Marx's thought emerged at a unique point in history during which many of the elements of the Communication Metaphor suddenly had a more central and a salient role than in the previous, less industrialized world. We have already seen how the development of technologies during the Industrial Revolution, such as transportation systems (the railway) and telecommunication technologies (the telegraph and the telephone), changed how the word "communication" is used and understood (see chapter 2). This is because the origins of the Communication Metaphor and of the Industrial Revolution are intimately linked. A Marxist-informed critique usefully fuses together key themes already established in this chapter: rationality, ideology, technology, expediency, efficiency, and productivity.

In effect, the contemporary Marxist-informed critique of capitalism is the milieu within which Katz's ethic of expediency is nurtured. Just as Marx understood the rationality of capitalism as an efficient and necessary system of production, twentieth-century theorists have more fully articulated the successes of instrumentalism, or technological rationality:[55]

> We are again confronted with one of the most vexing aspects of advanced industrial civilisation: *the rational character* of its irrationality. Its productivity and efficiency, its capacity to increase and spread comforts, to turn waste into need, and destruction into construction.[56]

Technological rationalism is the idea that rational decisions to incorporate technological advances into society can, once technology is ubiquitous, change what is considered rational within that society. For example, consider how the invention of the internet has had both benefits and costs as it has become essential infrastructure for the activities of everyday life. More specifically, consider how it has fundamentally changed our expectations for the nature of, and the pace of, our work and social relationships. Overall, we expect more and faster—more work in less time and more friends created faster and with less effort—but never mind the impacts on the environment or the epidemic of loneliness.[57] And lest you fear that these changes in expectations are individual preferences, be assured that they are predictable given the closed loop logic of the ethic of expediency: a technological rationalism enabled and justified by the ethic of expediency. In other words, while rational thinking as defined by science and technology can at once do *a lot* of good for society, at the same time it consigns to an often invisible periphery the collateral damage of environmental impacts, rising inequality, and individual suffering. Finally, by returning to the metaphor for technical writing as a coin or currency of scientific and technological progress, we are reminded that technical communication is a tool freighted with an ideology that "assents to its instrumental subordination to capital."[58] In this twentieth-century critique we hear the echoes of Bacon, Locke, Royal Society Plain Style, "severe reason," and the ethic of expediency reverberating across the centuries.

The notion of technological rationality as an ideology that shapes language practices might seem to be very far removed from the everyday lived experience of people who are trained in and work in scientific and technical disciplines. In part, this is because ideologies[59] do not comment on the unique, private thoughts and activities of individuals but work at the level of society and systems. Yet, ideologies do shape language practices that can become normalized and conventional for a particular community. One term that has been used to account for conventional scientific and technical writing practices informed by instrumental or technological rationality is "hyperpragmatism."[60] Hyperpragmatism is an ideology and set of practices that privileges utilitarian efficiency and effectiveness at the expense of sustained reflection, critique, or ethical action.

As an ideology that shapes cultural norms and practices in industry and government, hyperpragmatism also shapes the style and purposes of technical writing. We don't, of course, expect that a technical text would also perform a Marxist critique, or aim to please us aesthetically with beautiful language—the point is that the stylistic conventions for technical writing, such as objectivity and conciseness, are outcomes of socially shaped values and ideologies rather than the only way things can be—they are not a priori, or natural. For example, the dominance of the Royal Society's Plain Style for science writing emerged from a period of controversy about how science writing should be—Plain Style was not a foregone conclusion until it successfully became the only acceptable style for participating in institutionalized science.

A ubiquitous and traditional type of document that exemplifies hyperpragmatism is the software manual or help text. In a definition of software documentation from a textbook on how to write it, we can see the ethic of expediency explicitly at work: "Software documentation is a form of writing for both print and online media that supports the *efficient* and *effective* use of software for this intended environment."[61] A simplified but prototypical example of software document text might look like this—it should feel and look familiar:

How to Set Firefox as Your Default Browser

1. In the menu bar at the top of the screen, click **Firefox** and select **Preferences.**

   [image of Firefox menu]

2. In the **General** panel, click the **Make Default** button.

   [image of General panel]
   Note: If Firefox is already your default browser, the button will be missing and you will see the message, "Firefox is currently your default browser."

3. Close the **Settings** page. Any changes you've made will automatically be saved.[62]

In its aims, this short piece of help text has only one purpose—to support the web browser user in completing one very specific task—to set Firefox as the default browser—and nothing more. This text is pragmatic rather than reflective or critical. In no way does the text give the user reasons to reflect on whether it is a good idea to set Firefox as the default browser or what the consequences, both good and bad, individual or systemic, might be. While this kind of critical text might exist elsewhere—such as in a user forum—this help text, by its omission of certain types of information, assumes that, at best, reflection on decision making has already been completed, or, at worst, is unnecessary. This is to say that deciding to set a default browser and executing the task are considered separate activities with separate types of documentation to support them. The fact that we see nothing surprising about this state of affairs is an illustration of how normalized hyperpragmatism is in software documentation.

For would-be software documentation writers, the handbook explicitly promises that they will learn how to craft text that achieves the aims of efficiency and productivity of the software users:

> This chapter helps software documentation writers achieve two goals: encourage users to learn the program (proficiency) and encourage users to apply the program to problems in the workplace (efficiency).[63]

These statements about the purpose of software documentation are, indeed, hyperpragmatic in how they are structured around the ends defined by the need to increase worker efficiency and productivity (proficiency) when working with the software. Technical communication scholar Jennifer H. Maher argues that software documentation is shaped by and supports this hyperpragmatism because it serves the computer industry, and the capitalistic enterprise more generally, as *tools* to increase worker output.[64] Here again, we see the metaphor of technical writing as the coin or currency of technical and scientific work in a capitalistic industrial context. In this sense hyperpragmatism is an ideology, shaped by technological rationality, that determines the purposes and conventions for technical writing—hyperpragmatism is essentially how the expediency of deliberative rhetoric is played out in the context of technical work.

Technological Rationality Gone Wild: Revisiting the Transmission Model for Communication and Post–World War II America

> The postwar fallout of information theory is still with us.
>
> —Peters, *Speaking into the Air*

The "fallout" that J. D. Peters, a historian of communication, refers to is yet another echo of the yoking of technological rationality to scientific and technical language via the morally justified outcomes of productivity and efficiency. The final discussion in this chapter brings us historically into post–World War II America, an era we visited briefly in chapter 3 with a discussion of the transmission model of communication (fig. 4.2), a model born of post–World War II enthusiasm for understanding and solving very

120 / Busting the Myth of the Communication Metaphor

human problems via technological and scientific theory and apparatus. In chapter 3, I argued for the transmission model as an element of the Communication Metaphor because it is the fulfillment of the container, conduit, and windowpane metaphors as rationalized by its application to telecommunications technology. According to Peters, the transmission model (1948, 1949) presented a pervasive and seductive synthesis of the ethos of the times with the need for theories to explain the relationships between technological developments and human needs and capabilities.

This chapter has already traced multiple historical eras where the yoking of science, technology, rationality, and scientific and technical language has been normalized—starting with Bacon, then Locke, the Royal Society, the Industrial Revolution, the World War II military industrial complex, and software documentation. While Shannon's transmission model for communication may have originated in the mathematics of optimizing electronic signal transmission, it ended up, thanks to Weaver's generalization of communication, including "all the procedures by which one mind may affect another,"[65] reigning over the everyday realms of work, science, industry, education, government, and bureaucracy. And even though language and communication scholars have long moved on from this model because of how it excludes salient aspects of language and social theory that account for the problems of interpretation and power dynamics in language communities—essentially everything human—the transmission model still endures in a day-to-day sense. Figure 4.2 looks a lot like a diagram that

Figure 4.2. The transmission model: Shannon's graphical interpretation of his mathematical model for a communication system. *Source*: Claude Shannon and Warren Weaver, *The Mathematical Theory of Communication* (University of Illinois Press, 1949), 98.

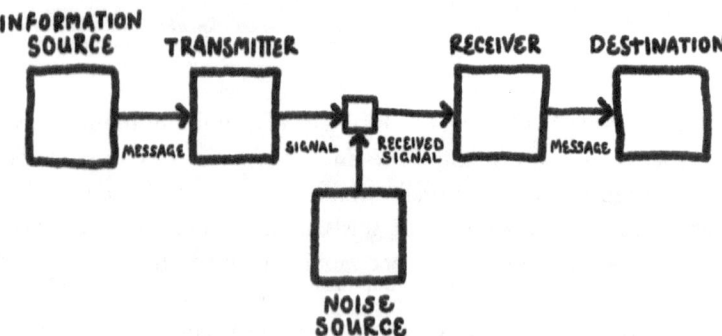

nearly anyone would sketch if asked to explain how scientific and technical communication works—just ask an introductory technical writing instructor what students draw when asked to define technical communication. The transmission model endures, I imagine, because at some level it just makes some common sense, experientially and conceptually, for a lot of people.

But why did the transmission model of communication become so dominant? Why not an alternative model that is more inclusive of the subtle power plays, intertextuality, identity negotiation, cultural diplomacy, flattery, and ego jostling that is natural to all human interaction, whether we admit it or not? Human communication, including technical and scientific communication, has never been only about bloodless information transfer—and, as this chapter argues, it has only been since about 1660 that we have believed in any sense that it is. According to Peters, the transmission model became entrenched in the post–World War II milieux because new fields of study, such as information theory, were burgeoning and promising new levels of achievement in how technology could improve society. Peters argues that Shannon's transmission model (1948) for electronic signal transmission, which Weaver essentially equated to human communication (1949), gave "American intellectual life a vocabulary well suited to the country's newly confirmed status as a military and political world leader." While initially what became "communication theory" was explicitly a theory of only *signals* and not of *significance*, or meaning, as the terms diffused through intellectual life—and they did at "violent speed"[66]—the distinction disappeared. During this era, *Information*, which is, notably, a nominalization, became a commodity with economic value. "Indeed," Peters argues, "the theory may have seemed so exciting because it made communication, something already quite familiar in war, bureaucracy, and everyday life, into a concept of science and technology. Information was no longer raw data, military logistics, or phone numbers; it was the principle of intelligibility."[67] In short, communication came to be understood in terms of the successful transmission of the signal, rather than the successful transmission of meaning.

The problem with the transmission model should now be a familiar theme of this chapter: It pretty much leaves out everything about being human. Just as a scientific and technical style ordered and shaped by expediency decenters human concerns via the morally justified aims of efficiency and productivity, so does the transmission model leave out the carnival of life by reducing it in communication to a signal problem: noise. In other words, it is a mistake, Peters argues, to assume that "better wiring will eliminate the ghosts."[68]

But here's the catch: In scientific and technical communication, better wiring has, in effect, eliminated the ghosts. In fact, I argue that the Communication Metaphor has eliminated the ghosts, when the ghosts are the voices that have been historically excluded from science and technical fields. The ghosts are the voices that have been excluded stylistically due to the rise to dominance of scientific Plain Style and the communities that have been excluded sociologically, because they have not had access to the institutions where scientific Plain Style, English Scientific Register, or the often-unacknowledged standards for written English, such as Standard American English, Academic English, or Mainstream American English are taught and normalized as everyday language.

Conclusion

If there is one upshot of this chapter it is this: IF as members of the scientific and technical community (including the author of this book!) we accept broadly that successful scientific and technical writing is "short and to the point, with facts only, no opinions," then it means that when we set out to write in scientific or technical style, we have, likely not consciously, already set aside human concerns, privileged the expedient, and reasoned within technological rationality and are engaged in hyperpragmatic aims. To be clear, rarely are these intentional choices—this is very definitely not a book about individual morality or ethics. But we are complicit in setting aside human concerns simply by accepting the style conventions that we have inherited. While we are not bad or unethical people for writing within the conventions of the scientific and technical community, acknowledgement of responsibility does force the recognition that technical and scientific style does not have neutral aims, that it cannot be used without complicity in the ethic of expediency, and that it should be handled like a dangerous tool—like a technology that can be applied to do good for the community but that, at the same time, enables harm. Technical and scientific writers desiring to escape the fraught personal and political entanglements more commonly thought to apply to political essay or creative writing might be disappointed that technical writing is not the refuge that they thought it was or that was presented to them. To this I can only offer a hearty welcome to the club of all writers who wrangle language to achieve a bewildering array of aims, while also trying their damndest not to cause harm.

Chapter 5

How Is the Communication Metaphor Perpetuated and Maintained?

Knowing now a little bit more about how to recognize the Communication Metaphor (chapter 2), what it means (chapter 3), and where it comes from (chapter 4), it's time to ask how the Communication Metaphor is perpetuated and maintained—why it sticks around so persistently. As this book has argued so far, the notion of communication that informs how we understand the purpose and style of technical and scientific writing is actually quite narrow—even to the point of being dogmatic. Conventional notions of technical communication are overwhelmingly shaped by the conduit, container, and windowpane metaphors and the transmission model for writing, the aims of efficiency and productivity, and seem overly obsessed with duplicating the contents of individuals' minds across time and space. As a reminder, here are a few of the collected comments from scientists and technical professionals about what makes technical writing successful:

1. Clarity is the most important element of successful written communication.

2. Successful written communication is:

    - concise, accurate, elegant, informative and engaging

    - accurate, relevant, easy to understand and accessible

    - short and to the point, with the facts only, no opinions

Of course, these answers are not news because we would fully expect them—in fact, it's very difficult to imagine what alternative responses might be. However, as I've argued, it's that very difficulty that is the germ of the Communication Metaphor. How is it possible to have such broad agreement, down to even the choice of words (e.g., *clarity*) for talking about a messy human process as complex as communication? Something out there is perpetuating and maintaining the Communication Metaphor. Part of that something is the beliefs about language and how language works that are considered conventional in contexts of education, industry, and government.

As I explained in chapter 1, my own awe at the pervasiveness, the invisibility, and the persistence of the Communication Metaphor stems from the predictability and conventionality of technical writing students' and technical professionals' answers when asked what makes technical writing successful. It also stems from the strong headwinds I encounter in technical writing classrooms when trying to turn the *Titanic* of student expectations for what technical writing is about: that technical writing, as the practice of transmitting technical information from one mind to another, is "short and to the point, with facts only, no opinions."[1] It's long been interesting to me as a technical writing teacher that most students come to the technical writing classroom with these beliefs about technical writing *already* firmly in place—this means that technical writing courses are not primarily a source of these beliefs but a reinforcer and perpetuator of them. As a site where the Communication Metaphor is explicitly perpetuated and maintained, the teaching of technical writing is a fruitful site for looking closely at these mechanisms of reproduction.

One of the beliefs about learning to write that students bring with them into the technical writing classroom—and usually also take away with them—is what linguist Laura Greenfield calls the Standard English fairy fale.[2] The Standard English fairy fale is the belief that there exists a pure form of English (Standard English) that is the most proper, and the most efficient—or clear—language for conducting the business of school, industry, and government. This chapter will begin with a discussion of the Standard English fairy fale and the harm that it promulgates as another element of the Communication Metaphor. It then continues with how the Standard English fairy fale informs how technical writing is presented in, and perpetuated by, writing handbooks and textbooks. The end of the chapter offers a glimpse into how technical writing teachers are trying to do best by their students by adopting a curriculum that balances the practical need to learn technical writing conventions with knowledge of and agency over the Communication Metaphor.

Readers of this book may come to this chapter with different relationships to the teaching of technical writing. Overall, this chapter is written for readers who are not scholars of teaching writing but who are either current students or who have taken a technical or professional writing course in the recent or deep past. Readers who are deeply versed in the scholarship of teaching writing may find the development of the Standard English fairy tale to be a bit detailed, but for those readers I offer near the end of the chapter a critique of how a common approach to teaching technical writing—teaching audience awareness—also, surprisingly, perpetuates and maintains the Communication Metaphor. Readers in industry may find parts of this chapter challenging due to the length and detail of some of the critique, but for those readers I encourage you to read chapter 5 as a setup to chapter 6, in which I imaginatively sketch what technical writing might look like in a post–Communication Metaphor world.

The Linguistic Facts of Life

If linguists proverbially ran the world, we might live in a world without the Communication Metaphor. This is because what linguists broadly accept as the "Linguistic Facts of Life"[3] disrupt the most powerful assumptions that maintain the Communication Metaphor. In summary, most linguists agree that:

1. All spoken language changes over time.
2. All spoken languages are equal in linguistic terms.
3. Grammaticality and communicative effectiveness are distinct and independent issues.
4. Written language and spoken language are historically, structurally, and functionally fundamentally different creatures.
5. Variation is intrinsic to all spoken language at every level.[4]

Yet, many non-linguists would beg to quibble, at times passionately, with some or all of these facts. As a student presented with these "facts of life" at the beginning of a first-year seminar on language diversity wisely raised her hand to ask, "If all linguists are in agreement about these phenomena, why is it that most people in general don't know about them or disagree about them?"[5] One answer is that linguists and non-linguists claim different kinds and sources of authority to "validate"[6] their beliefs about language. Linguists work from empirical evidence and draw conclusions based on accepted research methodologies. Non-linguists work from their own experiences with and stories about language, which are heavily influenced by how their worldviews and experiences are shaped by class, race, ethnicity, and their access to education. Between the facts of linguists and the beliefs and practices of non-linguists yawns a substantial gap. I believe that in this gap is a lot of information about the Communication Metaphor and how it maintains and perpetuates itself.

One powerful story told and retold in the gap is the Standard English fairy tale.[7] The Standard English fairy tale, which I include as another major element of the Communication Metaphor, is told by most people in the United States. It goes like this: Once upon a time there was a language, called Standard English, that is the most proper, correct, sophisticated, and clear way to speak and write English. It is the pure form of English and the natural choice as a common denominator language across diverse groups. We will all ride into the sunset together when *everyone* effectively speaks and writes Standard English.

As with the retelling of most fairy tales, a superficial understanding of the Standard English fairy tale focuses on how beautiful the princess Standard English is and how she got the handsome prince of lingua franca (the common language), while completely omitting the darker forces that are the fallout of the princess's success. This fallout is the harm that the Standard English fairy tale has caused through the creation, by definition alone, of the existence of nonstandard language practices and their resulting exclusion from use in education, government, and industry. A post–Communication Metaphor world will have to disrupt this fairy tale.

The plot of the fairy tale may be initially uncomfortable because it pushes up against the Linguistic Fact of Life that non-linguists often most vociferously disagree with: *All spoken languages are equal in linguistic terms.* As Greenfield argues in her essay "The Standard English Fairy Tale," this is because "most people in the United States believe that 'Standard English' is the most proper, sophisticated and clear way to speak English."[8] In addition, Greenfield argues that, "the language varieties deemed inferior in the United States tend to be the languages whose origins can be traced to periods in American history when communities of racially oppressed people used these languages to enact agency."[9] Greenfield points to American Englishes, such as Ebonics or African American Vernacular English (AAVE), Black English, Hawaiian Creole English, and Chicano English, as examples that linguists have documented to have logical, consistent rules of grammar but that have been marginalized as nonstandard. The exclusion of these Englishes from

Standard English has been motivated by standardized language ideology that idealizes one perfect *homogenous* language—the princess in the fairy tale. In the United States this homogenous—we might even say "pure," to return to a term that had such resonance in chapter 4—is, and no surprise here, the English of the disproportionately white middle and upper classes who maintain a hold on the majority of economic wealth and power.

Because this is a book about scientific and technical writing, not spoken language or language use in general, let's start by looking more closely at how the gap between linguists' and non-linguists' understanding of the two Linguistic Facts of Life that are relevant to written language perpetuate the Standard English fairy tale.

1. GRAMMATICALITY AND COMMUNICATIVE *EFFECTIVENESS* ARE DISTINCT AND INDEPENDENT ISSUES.

This statement means that just because a statement is said or written within the grammatical rules of a language, it is not a guarantee that the desired communicative aim will be fulfilled; in other words, because a statement is grammatical does not mean it is socially meaningful or appropriate. This understanding, however, just scratches the surface of the profundity of this "fact of life" for linguists: the relationship between grammaticality and effectiveness is actually quite complex, because linguists and non-linguists are likely to understand each concept differently:

Table 5.1. How Linguists and Non-Linguists Understand the Relationship between Grammaticality and Effectiveness Differently

	Grammaticality	**Effectiveness**
Linguists	A statement adheres to the internal rules of a language; it is intelligible to speakers or writers of that language.	The statement must be socially appropriate to the situation; *in addition*, the speaker/writer must be recognized as credible and authoritative to deliver the statement.
Non-Linguists	A statement adheres to an external standard, such as a standard language or style that is recognized as correct or conventional.	If the statement adheres to the right standard, the statement will be effective for all audiences who participate in that standard, regardless of who is speaking or writing.

In the difference between how linguists and non-linguists understand the relationship between grammaticality and effectiveness, we can see the Communication Metaphor at work. For example, a non-linguist's notion of grammaticality—that adherence to a standard language will ensure communicative success for all participants in that standard, no matter what the content is—relies heavily on the windowpane metaphor discussed in chapter 3. That is, language is a window that a reader looks through to see the content—if the window is "clear," then effectiveness is assured. In addition, effectiveness is understood to be ultimately dependent on the right use of language, not the complex social situation of the language use, including the social position of the speaker or writer in the larger context. In other words, linguists recognize that when it comes to effectiveness it matters just as much *who* is speaking or writing as *how* they are delivering the message. We'll see very soon how the linguist's view makes visible how the widespread expectation of standardized language use perpetuates the marginalization of speakers and writers for whom the standard language is not their home language, *even when* their language use conforms to the standard.

2. WRITTEN LANGUAGE AND SPOKEN LANGUAGE ARE HISTORICALLY, STRUCTURALLY, AND FUNCTIONALLY FUNDAMENTALLY DIFFERENT CREATURES.

This is an important point, as this is a book about technical and scientific *writing*, not spoken language. The differences between spoken and written language are both substantial and consequential, although, importantly, both spoken *and* written language do change and evolve over time. Differences include that spoken language is an innate human capacity, whereas written language is a skill that is taught, largely via formal education, with varying degrees of success. In a face-to-face spoken conversation, confusion and ambiguity can be resolved in real time via linguistic and contextual cues, body language, and changes in tone. In contrast, a written document enjoys no such opportunity for real-time negotiation and reconciliation with a reader. In addition, whereas a conversation is a temporal event that happens within a social context that informs its structure and meaning, a written document exists in perpetuity—it can travel far outside of the context that created it and exist long after the situation has ended. As a result, a written document relies heavily for communicative effectiveness on the ability of the reader to interpret it. Spoken language is fundamentally rooted in our concrete, communal experience of daily human life—our personal, work, school, and public lives demand it.[10] In contrast, the high value placed on written language is an outcome of the centrality to contemporary life of abstract phenomena such as education, the economy, history, the future, science, and government. With some exceptions, such as closed captioning and texting, when written language explicitly substitutes for spoken language, written language is rarely the province of the right now. Finally, spoken language evolves and changes at the speed of the cultural or historical change in human communities. Written language tends to evolve and change more slowly than spoken language, but it does change, and actually needs to. Standardization is more important to written language, which is more often read and interpreted by varied audiences in situations distant from the contexts and authors that created it—as a result, written language is more dependent on conventions and patterns for comprehensibility and effectiveness.

To the extent that there is ever slippage between spoken and written language in this book's discussions points to how porous the boundaries between the two can be in thinking about language, even though linguists are *very* clear that they are distinct. It is especially difficult for speakers and writers of what linguists refer to as Standard American English (SAE), Academic English (AE), Standard Written English (SWE), and Mainstream American English (MAE). In other words, when how you speak at home and how you are taught to write at school are more similar than different, it is difficult to believe that the differences are meaningful. For writers like myself, who learned a form of Standard English growing up in my white, middle-class home with two college-educated parents, the differences between how I spoke at home and how I wrote at school were nominal. I experienced from an early age that when I put my thoughts down on paper as I spoke them out loud or in my head, I produced a form of writing that would have authority and credibility in school and other public venues. In addition, being white aligned the standard English I spoke and wrote with my racial identity, which afforded me the credibility and authority to have a voice in settings that demanded Standard English.

In contrast, non-white speakers and writers of SAE or SWE, or what linguistic justice scholars[11] refer to as White English Vernacular (WEV) or White Mainstream English (WME) to acknowledge that what is standard has been set and enforced by a dominant racial group may not always have the same experience. Non-white speakers and writers may find that even when using the correct Standard English their voice is not always heard or afforded full credibility. Remembering that linguists include the identity of the speaker or writer as a factor of communicative effectiveness (table 5.1), the alignment of my race and my experience with language at school was not a coincidence but an outcome of the privileges that I inherited. My experience with the similarities between spoken and written language tells a story about the advantages, or privileges, I enjoyed growing up with such a thin veil between home, school, and public languages (or discourses), rather than offer linguistic evidence for the similarities between spoken and written language in general.

My experience of the continuity between spoken and written language may not be the same as speakers and writers of other Englishes, such as Ebonics, or African American Vernacular English—or Black English—Hawaiian Creole English, and Chicano English, which have have been historically and systematically labeled as vernacular or nonstandard.[12] Even though linguists[13] have shown through empirical research that these Englishes are as stable and rule-governed in both their spoken and written forms as SAE or SWE, they have had a socially constructed lower value, especially in school and other public settings. As a result, speakers and writers of these Englishes often don't have the same seamless experience of language use between school, work, and home language use as speakers and writers of White Mainstream English or Standard Written English. Put more plainly, the Standard English fairy tale excludes Black Englishes and other Englishes that have been historically marginalized from the Standard English of most school houses and workplaces.

It matters that we recognize the differences between spoken and written language because it clarifies one of the sources of harm perpetuated by the Communication Metaphor—that, in general, speakers of Standard American (white) English can't see or understand how much is at stake for students, workers, and citizens whose home language (or discourse) is not Standard American English to learn how to speak and write with authority and credibility in school, government, and other public places. And it's not just more labor-intensive but personally, economically, and culturally costly and, fundamentally, unjust. It is unjust that speakers and writers of Englishes that have been systematically excluded from Standard Written (white) English have to overcome additional barriers to participate fully in school, industry, and government. Even more so, it is unjust that the ways of thinking, being, acting, speaking, and writing expressed via so-called nonstandard Englishes are excluded from official public and economic life. This is an issue that decades of research on teaching speech and writing in school has addressed both theoretically and practically, but not always adequately. The main point here, however, is to recognize how difficult linguistic injustice can be to recognize if your own experience moving between speaking and writing at home and at school and work has been supported by the fact that those modalities are, for you, more similar than different.

Standard English Is Not a Language

At the core of the Standard English fairy tale lurks a dark secret: Standard English is not actually a language at all—it's more of a technology. Despite all of the potential for passionate positions on the necessity, even morality, of teaching a Standard English in school, in the end the debate is about not a language but a technology that has been designed, documented, manufactured, and disseminated for a purpose. While linguists can empirically document the unique grammars and vocabularies of all human languages as they are used by speakers and, often, but not always, writers (not all languages are written), the same is not true for a standard language, in particular Standard English. Standard English is formally taught and prescriptive (a designed set of rules for how language should be) rather than descriptive (how natural language is actually used by speakers and writers). Linguists cannot document standard languages because they don't exist outside of their inscription in handbooks, textbooks, teaching practices, and cultural mores—they don't exist in the wild. Historically, standardized languages have been invented and regulated by the people with the most economic, social, and political power as products of what linguists call the standard language ideology. Standard language ideology "proposes that an idealized nation-state has one perfect, homogeneous language."[14] While this ideology may not demand that all citizens speak the same standardized language in the privacy of their homes, it is necessary for all citizens to speak, and *especially* write, the same standardized language at school, work, and in government contexts. The fairy tale goes that everyone can, and should, learn the same standardized language technology to participate in the collective endeavor of efficiently managing and controlling the complex systems of society, such as education, government, and industry.

What standard language ideology means is that it's actually a lie to say that students are learning a new language as a lingua franca, or common language, to participate in society at large. What they are in fact learning is a set of socially and structurally—NOT linguistically—enforced rules for how to speak and write appropriately in a world where only White English Vernacular, White Mainstream English, or Standard Written English (or Academic English in educational settings) are acknowledged as standard.

If it's hard to believe that a standard English doesn't exist as a language, especially since many Americans, myself included, experience themselves as both speakers and writers of what is accepted as Standard English, consider the following results from a body of empirical studies that asked non-linguists about where "they speak correct English."[15] Based on these results, "Standard US English is the language spoken and written by persons"

1. with no regional accent;
2. who reside in the Midwest, Far West, or perhaps some parts of the Northeast (but never in the South);
3. with more than average or superior education;
4. who are themselves educators or broadcasters;
5. who pay attention to speech, and are not sloppy in terms of pronunciation or grammar;
6. who are easily understood by all;
7. and who enter into a consensus of other individuals like themselves about what is proper in language.[16]

I include these seven points to, I hope, make you chuckle. In other words, the number of people who would dare identify with all seven of the criteria could be anywhere from zero to hundreds to maybe thousands, to be generous—a number far from the imagined ideal of a language that is broadly intelligible across most of society! However, the myth of a standardized language is that it is generally intelligible, that it serves society as a whole and to be fluent in it grants access to the institutions of education, government, and industry. Never mind the irony that it is these very same institutions of education, government, and industry that are also maintaining and regulating Standard English the most rigorously! Whether or not so-called fluency in a nonexistent, gatekeeping language can, on its own, actually grant access to the educational and economic institutions of society is currently up for heavy debate, as we'll see below in the discussion about the justifications and approaches to teaching technical and scientific writing at the college level.

Standard language ideology also creates another problem: constructing a category for defining what is standard also definitionally creates the possibility of its opposite, nonstandard English, or everything the standard category excludes. The creation of the nonstandard category is not just semantics but has material consequences for the speakers and writers of Englishes that fall into this category. For example, a common student handbook for technical writing explicitly differentiates between standard and nonstandard English. The handbook entry "English, varieties of" doesn't only call into being the binary of standard and nonstandard English; it goes a step further to not only assign *who* uses each type of English but also explicitly assign positive and negative *value* to each kind:

> Written English includes two broad categories: standard and nonstandard. Standard English is used in business, industry, government, education, and all professions. It has rigorous and precise criteria for capitalization, punctuation, spelling, and usage. Nonstandard English does not conform to such criteria; it is often regional in origin, or it reflects the special usages of a particular ethnic or social group . . . Nonstandard English is characterized by inexact or inconsistent capitalization, punctuation, spelling, diction, and usage choices.[17]

The preface of this handbook of technical writing frames its purpose as within the project of equal opportunity: "The Handbook of Technical Writing is the text students and professionals need to land, navigate and stand out on the job."[18] However, the passage about standard and nonstandard English makes it pretty clear who needs this handbook: all readers of "particular ethnic or social groups" who are not already participants in Standard Written (white) English. Given the handbook's explicit embrace of the standard/nonstandard binary, it appears that the handbook is unconcerned with perpetuating the Standard English fairy tale with Snow White leading the way.

It's important to acknowledge that my point here is not to completely discredit the value of a standardized written language for facilitating communication across diverse communities. The alternative to an enforced Standard English is not an open field day to write engineering reports, scientific articles, or anything else any way that you want and then to blithely expect everyone to be able to understand you perfectly. Such a slippery-slope argument—one that calls on the most extreme alternative to discredit the original point—does nothing to productively enhance our engagement with the problem. Certainly, some common language practices and agreements are essential to facilitate communication across a diverse society—and written forms of languages cannot avoid them. No one, not even linguists, disputes this point. The problem is that what has been codified as Standard English has historically been the standard English of only part of the population: the disproportionately white middle and upper classes, to the exclusion of most others. In short, Standard English is not representative of all English speakers and writers, only a select few. The main point is to dislodge the mantle of fact from Standard English and Standard Written English.

Whether or not to believe that Standard English is a fairy tale remains the luxury of the non-linguist, since empirically the matter has been settled. To continue to believe, however, has troubling consequences for individuals:

Greenfield argues that believing and behaving as if Standard English exists is motivated by fundamentally racist beliefs. Greenfield calls to the carpet scholars and writing teachers, like myself, who, as honorary linguists, might feel superior to non-linguists in our acceptance of the linguistic equality of all languages. She points out, however, how this superiority actually perpetuates racist assumptions about writing. She argues that when we trumpet language diversity as a value in our writing assignments and curriculums, we are accepting what she calls the "most insidious and false assumption"[19] that a standard English actually exists *within* the diversity of all Englishes and other languages. If teaching writing courses for language diversity means recognizing and celebrating students' unacknowledged Englishes or other languages *alongside* or *mixed into* (linguists call this code-switching) Academic English or Standard Written English, we are simply showing our tolerance for how nonstandard language use disrupts the "purity" of a homogenous standard English without actually troubling the notion of the existence of the "standard" or decentering its power to shape what and how we teach or how our students write. Fundamentally, the problem begins with our acceptance of the existence of a standard at all.

It is important to ask at this point "What is wrong with asking everyone to learn a common set of writing conventions for the good of all?" Common sense argues for it, so, the first answer, of course, is nothing. But everything depends on what those conventions are and whether they impede or enable linguistic justice. The previous section discussed the privilege inherent to growing up speaking and writing Standard English in the home and the community and the resulting blindness to linguistic injustice. In addition, the belief in the *benevolent* nature of teaching a standardized language that *just happens to be white* renders invisible to many the harm perpetuated by the exclusion of Englishes of historically marginalized communities. It is a core belief of non-linguists, and a central plot in Greenfields' Standard English fairy tale, that common sense and a practical expediency render it common sensical to label some Englishes as nonstandard while promoting the English of only some communities. This core belief can be hard to recognize, because, like all elements of the Communication Metaphor, the Standard English fairy tale works for a lot of people most of the time, yet it also excludes some people *all* of the time.

But an alternative is possible, although at this point it requires imagination. If we accept that Standard English, whether spoken or written, is not a language but a technology, because it is not a quantifiable dialect with a finite set of rules and features, *and* we acknowledge that there are multiple dynamic Englishes with equal value, *and* we assign no innate superiority to those inflected as white, then we open up a possibility for a world where we might achieve something closer to linguistic justice. In this world, and in particular in scientific and technical contexts, we would purposefully define and teach Standard English to be a bigger, more dynamic tent that valued including, rather than excluding, the rich variety of Englishes that communities use to make meaning. Importantly, placing a high value on including a rich diversity of Englishes in writing wouldn't be limited to the authorship of individuals writing personal essays but would be integral to writing for school, government, and industry more broadly. Over time, the category of Standard English might erode because the category of nonstandard would, in the abstract as well as in practice, become null and void—at least in the arenas of writing that are shared across society, such as scientific and technical contexts. As a result, everyone would benefit: Both written and spoken English would afford more ways of thinking and knowing that are essential to addressing pressing and complex issues of science and technology, such as climate change, artificial intelligence, disease, political division, and much more. In addition, everyone would be challenged to learn ways of speaking, writing, and and thinking that go beyond how they speak and write in their home communities. Certainly, this alternative is no panacea for achieving linguist justice, but just because we cannot *yet* fully imagine how a world like this might actually work, particularly in technical and scientific contexts, is not a reason to reject the validity, and urgency, of the argument for considering it a possibility.

Handbooks: The Commodification of What Is Standard

If standard languages are technologies, then writing handbooks package and commodify the technology for general consumption. And writing handbooks are ubiquitous: a quick online search for "writing handbooks" will deliver thousands of results promising writers success across the diverse forms of writing, such as academic writing, research writing, legal writing, report writing in many specified medical and technical fields, grant writing, media writing, science writing in general and within specific fields, accounting writing, novel writing, comedy writing, poetry writing, social media and online content writing, and it goes on. Handbooks for writing, which straddle the spheres of education, the workplace, and the personal library or desk, have been central to the development of Standard Written English. A handbook is, by definition, a guidebook for how to do something, and writing handbooks have been guidebooks for how to write. Handbooks or guidebooks, as distinct from textbooks, are meant to support a writer during the act of writing, so they are organized to be quickly accessible and offer direct, immediately applicable, and actionable advice. Their accessibility, however, comes at the cost of any discussion of the history or the rationale behind the advice. The omission of complexity in the delivery of rules for language use in handbooks, as we'll see demonstrated below, is a key perpetrator of the Standard English fairy tale.

The writing handbook tradition in English has a complicated, centuries-long history. In sweeping terms, it emerged in the seventeenth century to support the burgeoning numbers of people entering into literate professions, such as law, business, and academic fields. The publication in 1762 of *A Short Introduction to English Grammar* by Robert Lowth, poet and bishop in the Church of England, formally initiated the tradition of regulating a standard written English via a published handbook that could be purchased. Many of Lowth's style prescriptions, such as avoiding the double negative and not dangling modifiers, have been nearly unanimously carried forward into modern writing handbooks.

Just as Lowth's writing handbook served to regularize writing in the professions, it also encouraged class climbers and social aspirants to conform their spoken and written language use to a prescribed correct form. This blurring of the norms for language use in professional spheres and the language norms of elite society persists today. Readers are likely familiar with Oliver Strunk and E. B. White's bestselling style guide *The Elements of Style* (first published in 1935, with subsequent editions in 1959, 1972, and 2000). This book is nearly an exact analog to Lowth's, especially in its purpose to help writers achieve a style of writing shaped by the early- and mid-twentieth-century social anxieties of the rising middle classes and the desire to reinforce social strata via written language use.[20] That *The Elements of Style* has also been one of the most often assigned grammar and style handbooks in college writing classes[21] is evidence for the systemic lack of distinction made in educational settings between Academic English and White Mainstream English or White English Vernacular.

Finally, the composition handbook tradition of the late nineteenth and early twentieth centuries[22] supported the teaching of academic writing in freshman English courses at universities. Because so many students took these classes, these composition handbooks significantly shaped public attitudes about what makes for good writing, with a general drive for a focus on correctness and prescriptiveness through most of the twentieth century. Finally, a related tradition for specialized handbooks for writing in scientific, engineering, and technical fields also emerged.

While histories exist for the various lineages of writing handbooks, I've not found one that has traced how common elements of the Communication

Metaphor, such as the values of clarity and objectivity and the moral imperative for grammatical and stylistic conformity and correctness, have traveled across the writing handbook traditions. Consider this question: If the Royal Society's Plain Style developed in the seventeenth century, how did this style shape the eighteenth-century initiative to formalize a standard form of written English across the professions and for educated society in general? In other words, when a scientist or technical professional today describes successful writing as "concise, accurate, elegant, informative, and engaging," to what extent are these values for scientific or technical writing the inheritance of the society's Plain Style, and to what extent are these values a product of what has been a multicentury campaign to standardize English across all of society, what is often referred to handbooks as "plain English"?[23] Across writing handbooks of all types, writing considered accessible to broad audiences is variously referred to as *Plain English, plain English, Plain Style, plain language*, and *Plain Language*.[24] Given the slippage in what each term means, it's really very difficult to say when one or the other is behind a statement about stylistic values. However, asking the question usefully foregrounds how little understanding we have about the sources and justifications for grammar and style conventions that are nearly universally accepted today, especially in technical contexts. Such a disorienting fog adds to the evidence for the Communication Metaphor as a complex of tacit assumptions and practices for writing, including the Standard English fairy tale, that are so widely accepted yet so little understood.

Modern handbooks for scientific and technical writing carry on Lowth's and Strunk and White's tradition of providing a finite set of approved rules with murky origins that can be directly applied to achieve successful written communication. A guidebook for science writing from 2013 comforts scientists and science students sitting down to write with the promise that they can avoid the "poor writing" that pervades most of the scientific literature:[25]

> The good news is that you can write science in plain English by applying a relatively short list of principles developed for professional writers by Joseph Williams in his book *Style: Toward Clarity and Grace*. These principles are based on linguistic theory about what readers look for when they read complex, unfamiliar information.[26]

We should be concerned by several of the claims made in this passage, especially since the Communication Metaphor announces itself forcefully in the word "clarity." Joseph William's book was originally published in 1981, with multiple editions since, but was intended as a writing handbook primarily for use in freshman college writing courses. The title's suggestion that clarity and grace, a highly subjective and morally if not religiously inflected stylistic goal for academic writing, points again to the blurring of boundaries between the values for elite social discourse and academic writing in writing handbooks. The author of the science writing handbook readily extends William's values to science writing without acknowledgement of (1) any differences between the aims of general academic writing and science writing, and (2) how these values, when distilled into fixed sets of rules validated by "linguistic theory," perpetuate the exclusion of many writers for whom an elite social discourse is not already a norm or who would experience, expressly or implicitly, the advice to conform to a set of fixed rules as a form of personal critique, exclusion, or even violence.

Beyond this example, the notion of clarity, one of the pillars of the Communication Metaphor and the key to the Standard English fairy tale (clear writing is standard writing), pervades technical and scientific writing handbooks and recalls the black-boxing and nominalization of communication as a messy process, as discussed in chapter 2. Consider these statements from readily available technical and scientific writing handbooks:

1. This book dispenses with information about what scientists write and focuses entirely on how to write clearly and comprehensibly.[27]

2. You cannot achieve your purpose or a goal like persuasion without clarity.[28]

3. Clarity is essential to effective communication with your readers.[29]

These statements about clarity, while apparently hopeful, at best promote unuseful and vague conventional wisdom, and at worst prop up the Standard English fairy tale. They each cast clarity as an abstraction, a metaphor, a rhetorical ideal, and even a doctrine for what counts as successful writing.

On closer inspection, each statement blatantly depends on multiple elements of the Communication Metaphor. Coding each statement according to the schema developed at the end of chapter 2 makes this dependance visible:

STATEMENT 1

Statement broken down by word or phrase	Element of the Communication Metaphor (code)
This book dispenses with information about what scientists write and focuses entirely on how to write clearly and comprehensively	Windowpane theory of language [content is separable from language] Windowpane theory Conduit metaphor, rhetorical framework

STATEMENT 2

Statement broken down by word or phrase	Element of the Communication Metaphor (code)
You cannot achieve your purpose or a goal like persuasion without clarity.	Rhetorical framework, expediency Rhetorical framework Windowpane theory

STATEMENT 3

Statement broken down by word or phrase	Element of the Communication Metaphor (code)
Clarity is essential to effective communication with your readers.	Windowpane theory Black box[1] Conduit metaphor, rhetorical framework

---

1. A new code not introduced in chapter 2 but a familiar concept. In the coding scheme a segment is coded as "black box" when the terms *communication* or *communicate* are used without definition or explanation. When the word is used in such a way that the reader must supply all the procedural and definitional knowledge for the term. For the full scheme, see S. Read, S. "100% Say Writing Is Important to Their Work, but What Harm Does This Uncontroversial Finding Obscure? Early Results from a Survey of Scientists and Technical Professionals about Writing and Communication," presented at IEEE Conference on Professional Communication, Limerick, Ireland, July 17–20, 2022.

While the codes make visible that the Communication Metaphor shapes these statements, a closer look at each one sheds additional light on how these statements are problematic. After a strong statement declaring that content, what scientists write about, is wholly separable from how they write, statement one calls out its own vagueness by the need to teach how to write both clearly *and* comprehensibly, as if one is differentiated from the other. This may seem like a small point, and it is, but if comprehensibility is a judgment of the audience, then clarity, to be something different, must be an independent characteristic of language, and therefore fully governable by a fixed set of rules for writing. This usage promotes the myth that a fixed set of stylistic of rules—or standardization—can consign language to neutral invisibility in order to let the truth of the message shine through for the audience. This is the classic windowpane theory of language, which, as we learned in chapter 3, perpetuates the false expectation that if only language could be made completely transparent, then knowledge would shine through and the reader would bask effortlessly in new knowledge.

Statement 2 links clarity directly to achieving the rhetorical goal of persuasion. Again, this statement is innocuous on the surface, but a closer inspection reveals how it is code for promoting the most exclusionary aspects of the Standard English fairy tale. To persuade an audience to accept a statement as true or to adopt a new or existing position on a topic requires that the reader is willing to assign authority to the text. This judgment is not just about the acceptability of the language but also includes an evaluation of the author's credibility, who may or may not be explicitly given a byline. As this statement suggests, language deemed not clear—or standard—is much less likely to be assigned credibility, so adherence to standard English is deemed necessary for communicative effectiveness.

Adherence to Standard English, however, also turns out not to be sufficient. Language judged as not clear may still persuade a reader if the author can be assigned authority. A classic example would be university students who are asked to read works by French philosopher Michel Foucault in their courses. Foucault's language and argumentation are notoriously elliptical and abstract and challenge every American's expectation that academic language and argumentation should be written for expediency and the delivery of logically ordered points. Yet, while students express frustration with the abstruse text, they largely assign the blame for their difficulty to their own inadequacies as readers, and preserve at every turn the authority of Foucault to speak on matters of philosophy. However, not all authors, especially those who are not old (now dead, since 1984) white males, enjoy such a benefit of the doubt from readers. A perfectly clear or standard piece of writing authored by a person of color, a native person, a non-native English speaker, or a person from another minoritized group may not automatically be granted full authority or credibility, despite—or because of—using standard language. This is a tricky point to not overstate, but it is a very important one. The point is that communicative effectiveness is not just about the stylistic success of the language; it is also about the reader's *judgment* of the author. The reader brings their own biases to the reading of text, and the lens of those biases extends to who the author is as well. Contrary to the confident statement that clarity in language results in achieving persuasive goals, communicative effectiveness depends just as much on the audience's willingness to listen.[30]

Statement 3, to keep it brief, recalls the tautology from chapter 2 that is the product of the black-boxed term *communication*: It is important to learn technical communication to learn skills to *communicate* more effectively. Except that in this case the equally capacious term clarity is equated with communication: *clarity* is essential to effective *communication*. Statement 3 is a prime example of a commonplace (familiar saying) whose truth is so self-evident that it can be assumed that the meaning the reader supplies for the two abstract terms (*clarity, communication*) will be highly conventional and therefore mutually comprehensible.

Factoid Advice with No Further Comment: Technical Writing Handbooks

The handbooks for technical, scientific, or professional writing retell the Standard English fairy tale in the context of writing for science and industry. As a longtime teacher of technical writing, I've always struggled with if and when to ask my students to buy a readily available technical writing handbook—in print or digital form—published by a large textbook publisher. Arguments for assigning the handbook include giving students who are completely new to technical writing an accessible resource to turn to for basic concepts, vocabulary and grammar, and style topics. The impulse here is primarily practical—to give students a useful tool to support their learning. The logic goes that if students will be judged in the real world on whether their writing is read as professional, clear, or credible, then they need to know what the broadly held rules are for achieving these rhetorical aims. Yet, to frame the handbook as a useful tool projects a kind of benign neutrality on Standard English that I already know is false. Greenfield's heavy critique of teaching all students a Standard English in the service of equity while turning a blind eye to who is excluded or harmed by this practice weighs heavily on my decision.

Additionally, I question the value of assigning a handbook due to my deep skepticism about whether students actually actively use a handbook while writing or editing. Do they just keep it on a shelf as a talisman to the project of learning how to write "well"? If you yourself own a writing handbook, when did you last consult it? Finally, while alphabetically organized handbook entries on topics of grammar and style offer quick and direct definitions and examples that can be quickly applied—what I consider factoids about style and grammar—entries omit any context for the real purpose or the historical origins of the style or grammar directive. The upshot is that if I assign a handbook for technical writing, then I am implicitly validating both the existence and the authority of the technical writing handbook, for what I fear is at best a minor practical gain for students. In short, I am playing my own part in uncritically perpetuating the Standard English fairy tale.

Let's look at an example of a handbook entry to understand how much information and context is omitted and to better understand why this matters. The entry on double negatives[31] in a commonly assigned technical writing handbook is an example of how factoid advice about a seemingly small issue of style actually conceals centuries of marginalization of Englishes that have been excluded from what is considered standard. As you read the entry text below, reflect on what it would feel like if the sentence labeled as "unclear" is one that you have heard recently spoken or written in your own community. For some readers I know I don't need to prompt reflection at all.

> Double Negative
> A double negative is the use of an additional negative word to reinforce an expression that is already negative. In writing and speech, avoid such constructions.
> Unclear: We don't have none. [This sentence literally means that we have some.]
> Clear: We have none.

This handbook entry assumes rather than explains the reasoning behind the directive. The offered example illustrates the conventional stylistic notion that in Standard English two negatives in a sentence equal a positive: to say or write "We don't have none" would actually mean "We have some." The claim that the double negative construction risks confusing the reader might initially appear value neutral. The claim is based on a kind of mathematical logic: a double negative in a clause construction is actually equal to a positive because the two negatives cancel each other out. In mathematics, some operations that involve two negative numbers, such as multiplication, do in fact produce a positive result. However, the plot thickens if we consider that "We don't have none" actually doesn't mean "We have some" to the communities of writers and speakers who would (standardly for them) use this construction. For speakers and writers of AAVE, the double negative emphasizes, rather than negates, the negative connotation: "We don't have none," in fact means both we "don't have" and we "have none." For the record, mathematical logic supports this understanding as well: when two negative numbers are added together, the result is a more negative number. Evidently, the double negative rule is not about a preference for multiplication or addition—mathematics is besides the point. So, what is the stylistic convention in Standard English against the double negative really about?

Unsurprisingly, it is not the cold logic of mathematics that has established the double negative as bad style. To recall Greenfield's pointed critique from above that it is the Englishes of Black and other minoritized communities that have been purposefully excluded from Standard English, it's not a surprise to learn that the double negative is a standard part of AAVE, and other Englishes as well. In fact, the double negative was a part of English for centuries before it became marginalized as nonstandard. Double negative constructions appeared in the works of Chaucer and Shakespeare as well as the first translation into English of the King James Bible (1611). Most histories of the double negative assign the origin of the stricture against it to the publication of Robert Lowth's *A Short Introduction to English Grammar* in 1762. Lowth's book was essentially the first handbook to regularize a standard English for writing, and many of his prescriptions have been unanimously carried forward into modern handbooks. On the double negative he wrote, "Two Negatives in English destroy one another, or are equivalent to an Affirmative."

Despite the persistence until today of Lowth's stricture against the double negative, historians debate Lowth's original purpose in banning the double negative construction. One theory is that Lowth aimed to create a prescriptive grammar to linguistically define an elite social class of which he was very much a member. Did Lowth purposefully aim to marginalize forms of English spoken and written by the lower classes to facilitate his own social ascension? A study[32] of Lowth's correspondence within his social network, which included the royal family, leaders in the Church of England, and other high society, tried to find evidence that Lowth's purpose in his prescriptive grammar in general, and in the case of the double negative in particular, was to reinforce the boundaries of social class via language use.

The plot thickens when we learn that Lowth's oft-cited stricture against the double negative did not appear until the second edition of the handbook was published in 1763.[33] This suggests that it may have been added after critical feedback noted the original omission. Lowth may not have originally seen a need to call out the double negative as nonstandard since the evidence of his letters suggest that among his social peers it had already fallen out of use among educated[34] writers. In fact, linguistic evidence suggests that by the sixteenth century the double negative had already become the social marker that it is today, and that the "social aspirers"[35] who led the eighteenth-century movement for a more regulated written English were already hypercorrecting for single negation.[36] The puzzle this presents historians is why Lowth's handbook would explicitly include a stricture against a construction that was no longer in use among the population who would ostensibly be using the handbook. Certainly, its inclusion did not have an entirely pragmatic purpose. Whether Lowth was personally behind the stricture against the double negative or not, its inclusion in the second edition confirms its power as a marker of social standing, which was already in full force when Lowth published the handbook. In general, the publication of Lowth's handbook initiated the formalization of a tradition of prescriptive "correctness" that still pervades writing handbooks today; in fact, his handbook reflected what was already a widely practiced social norm for sorting people by social class as marked by their language use.

Lowth and the eighteenth century may seem a long way away from an entry on the double negative in a 2019 edition of a technical writing handbook. Certainly, a handbook entry, with its reader-friendly purpose to expedite "clear and correct" writing for the professional or technical writer, cannot include all of this history. And that, right there, is the main problem. The problem is the whole idea of a handbook so concise and usable that it elides the complexities and histories of language use, even when these histories still *materially* affect people today, including some of the readers of this book. Can handbooks have a purpose beyond telling and retelling the Standard English fairy tale? That is the question.

Textbooks: Propagating the Conduit, Container, Windowpane, and Transmission Metaphors

If technical writing handbooks offer streamlined stylistic rules and advice that retell the Standard English fairy tale and perpetuate other elements of the Communication Metaphor such as correctness, their close cousins, technical writing textbooks, do the same with more words, and with a more holistic goal to teach the process and conventions of writing within the richer context of a semester-long course. After the first-year composition course meant to initiate new college students into writing in college, the second most common writing course at universities is the generalized professional and technical writing course intended to initiate students into writing in the sciences and the professions. At a rough estimate, forty thousand sections of introductory technical, scientific, and/or professional writing are taught at US universities every year.[37] If this is the case, roughly one million students learn introductory technical writing every year—this is not a small number. Readers of this book may recall taking a course similar to this, or may currently be taking such a course. Regardless, the curriculum and textbook resources presented in the introductory technical writing course have a significant influence on students' lifelong attitudes toward and practices of technical writing. Technical writing course textbooks, therefore, are another way that the Communication Metaphor is maintained and perpetuated. In general, technical writing textbooks are primers for the conventional language about technical writing that shows up in the survey of scientists and technical professionals—they are not the only sources of this language, but they are common and authoritative.

Standard, meaning commonly assigned and conventionally organized, technical writing textbooks unwittingly promote the central metaphors of the Communication Metaphor, with few exceptions. After all, to *not* rely on the conduit, container, windowpane, and transmission metaphors would require the work of this book to be complete! In most textbook writing, the container, conduit, and windowpane metaphors slip by as uncontroversial instructional statements. For example, in one of the most widely used standard technical writing textbooks, *Technical Communication* by Mike Markel and Stuart A. Selber, the container metaphor structures what we learn about the purpose of documents:

1. *In* each of these documents [journal article, press release, infographic, animated blog post], you present the key information.[38]

2. Generating ideas is a way to start mapping out the information you will need to include *in* the document.[39]

In the same textbook, documents can also be construed as conduits:

Technical information is frequently communicated *through* documents such as proposals, emails, reports, podcasts, computer help files, blogs and wikis.[40]

And sentences as well:

> Good technical communication consists of clear, graceful sentences that *convey* information economically.[41]

The Standard English fairy tale, discussed at length above in terms of handbooks, also lives on untroubled in most textbooks. Often it is cloaked as the neutral value of correctness, another element of the Communication Metaphor:

> A *correct* document is one that adheres to the conventions of grammar, punctuation, spelling, mechanics, and usage. Sometimes, *incorrect* writing can confuse readers or even make your writing inaccurate. The more typical problem, however, is that *incorrect* writing makes you look unprofessional. . . . If readers doubt your professionalism, they will be less likely to accept your conclusions or follow your recommendations.[42]

This quote, from the same textbook as the metaphor examples above, was offered in chapter 1 as evidence of how widely accepted and uncontroversial statements that rely on "correctness" as their measuring stick for successful writing may actually be thinly veiled racist sentiments sustained by the Standard English fairy tale and the product, ultimately, of historic, systemic, white supremacy. Certainly, editing and proofreading are essential steps in the preparation of any professional document; the problem is in the difficulty when pursuing correctness of differentiating between an error and nonstandard language use. I'm not confident that most technical and scientific writers would even be able to differentiate—for example, when is the lack of verb agreement an error and when is it nonstandard language use? Before becoming too settled on your own answer to this question, recognize that this question cuts to the very heart of the Standard English fairy tale. I include this quote here again in the hopes that, given the extensive analysis and discussion of the Communication Metaphor and, in this chapter, the Standard English fairy tale, it is now abundantly evident that they cause harm by purposefully excluding ways of using language and making knowledge and the people who belong to those language (discourse) communities.

I would hope that at this point in the book these textbook examples are unsurprising, even redundant. Therefore, we must start attending to what an alternative might look like. A few textbooks do explicitly challenge conventional metaphors for technical writing. These books are, unsurprisingly, written or cowritten by linguists! One of those textbooks is the *IEEE Guide* to *Writing in the Engineering and Technical Fields* (2017), a text that I have assigned in both general technical writing and science writing courses. This text is sophisticated in its presentation of how writing in technical contexts is *rhetorical* and *social,* meaning that even technical texts need to persuade audiences of their utility and authority, and that technical texts don't just *convey* information but participate in a community project of *making* knowledge. Overall, the *IEEE Guide* is savvy about situating the writer in a complex social environment that shapes how the writer will use language to achieve communicative effectiveness. It also explicitly resists the conduit and transmission metaphors for communication:

> Because all communication reflects human relations, many technical professionals acknowledge that writing is *more than a neutral conduit to convey* information from one person to another. Instead, engineers and technical professionals shape knowledge as it moves between the professional and a client or the end users of the technology.[43]

Notice how the authority for this sentiment that "writing is more than a neutral conduit" is actually attributed to "technical professionals," not the textbook authors! This is because, when technical writing students become technical professionals, they learn that even technical writing is messy and unreliable when it comes to achieving communicative effectiveness in the real world, especially when you have to write for multiple, often competing, audiences, such as your boss and the customer, at the same time. Ironically, in a Communication Metaphor–saturated world where successful technical writing is widely known as "short and to the point, with the facts only, no opinions," experienced technical professionals also know that an engineer who has the rhetorical savvy to accommodate the interests of multiple stakeholders is actually a stronger writer. For example, an engineer might have to make a careful choice in reporting results of a product test to a customer that both honors the desire for management to show the customer that the product will serve their needs and acknowledges that from an engineering perspective the product is not yet fully working.

For example, one engineer at an aerospace company[44] shared with a writing researcher how he struggled to express disappointing results of a product test in a report that, once approved by management, would go to the customer. While as an engineer he felt that one aspect of the product test had actually been "hopeless," he wrote the sentence to say that the test had been "fruitless initially" in order to convey a sense that it might at some future time work. He felt justified in his choice based not on his engineering knowledge but on his knowledge of what management would approve so that the product could move forward for everyone's benefit. In this context the engineer was recognized as a good writer because he showed that he could "negotiate successfully the subtle boundary between, on the one hand, the stylistic and formal demands of clarity, neutrality . . . and, on the other hand, the institutional, social and situational (read: political) demands placed on the text."[45] Testing reports written only from the engineer's perspective that didn't account for the perspective of management and the customers were often marked as "bad" writing and sent back for multiple revisions, costing the engineers and the company time (and therefore money).

Chapter 3 discussed in detail how the conduit, container, window pane, and transmission metaphors are all based on the myth that communication should be effortless. When it isn't, failure is often attributed to a lack of clarity, which could be an oblique reference to nonstandard language use. In addition, the lack of clarity is understood as a characteristic of the language rather than as an aspect of the relationship between the writer, the text, and the reader. The "fruitless initially" example contradicts this myth, because it shows how technical writers make writing decisions in response to the complex context, or rhetorical situation, that they are writing in.

It might seem that, given the success of the engineer to make writing choices based on the needs and expectations of both his manager and the customer, an effective way to teach students to think and write beyond the conduit and transmission metaphors is to teach technical writing students to think carefully about the audiences that they are writing to, for, and with and to adapt their writing choices accordingly. And, thankfully, many introductory technical writing textbooks, if not most, already do. While this is, in the main, a good thing, I'll still argue below that teaching students about audience awareness is not nearly sufficient to reject and replace the Communication Metaphor.

Teaching Rhetorical Thinking Is Not a Way Out of the Communication Metaphor

This section challenges another conventional dogma for writing successfully: know your audience. Scientific and technical writing textbooks don't just teach conventions for style and how to write "correctly." Textbooks also, rightly, aim to teach students how to be savvy writers, like the engineer above who struggled with how to turn "hopeless" into "fruitless initially."

Teaching rhetorical thinking, which is generally referred to as audience awareness or writing persuasively to defined audiences, is a pillar of most contemporary writing textbooks. Textbooks, however, rarely address the conflict inherent to teaching technical writing style as a neutral conduit while also encouraging writers to take into account the needs and views of audiences with varying priorities, such as managers and customers, or scientists and funders. In addition, textbooks tend to position technical writers as benign, even benevolent, helpers or advocates for their audiences, even if the writer is in a position of power with interests that conflict with those of the audience. The Communication Metaphor perpetuates and maintains this conflict, as we'll see below.

Most technical writing teachers would agree that as long as words, sentences, and documents are chosen and authored mindfully with the needs and desires of a very specific audience in mind—what we call rhetorical thinking or audience awareness—then the potential for harm is very low. Markel and Selber's textbook agrees that audience awareness is important:

1. When you communicate in the workplace, you have not only a clear purpose—what you want to achieve—but also a clearly *defined audience.*

2. One of the most challenging activities you will engage in as a professional is communicating your ideas to audiences.[46]

While examples 1 and 2 may seem like banal givens for any kind of writing, true rhetorical thinking is a trained skill that requires deep knowledge of the writing situation that goes beyond knowing superficial characteristics of the audience, such as age, gender, or level of technical training.

The textbook *Writing Science Right* (2018) begins confidently with "Chapter 1: Writing for the Readers: Know Your Audience and Analyze Their Needs."[47] The chapter opens by explaining to students that successfully writing science, or writing "to explain, describe, argue for or against scientific ideas," requires learning to write for a variety of readers that is considerate of the purpose for writing and the knowledge level of the readers. The advice seems straightforward: "The foundation of writing science and any technical writing is knowing how to write for its readers."[48] This is good advice: open, empathetic, listening, adaptable, and sincere. There is nothing controversial or wrong about this advice. It is practically gospel.

However, no matter how keenly felt or worthy a reader-centric approach to teaching technical writing is, we'll see on closer inspection that the narrow kind of rhetorical thinking and awareness that we develop in technical writing courses reinforces how the conduit and container metaphors exclude certain audiences and alternate theories of how language and knowledge work.

In *Writing Science Right*, the explanation of what it means to write for readers explicitly relies on the container and conduit metaphors from the get-go. After the introductory paragraph, a list of apparently unproblematic goals for the writer follows the opening paragraph:

1. Readers must understand the meaning of what is written exactly as the writer intended.
2. Writing must achieve its goal with the reader.
3. Writers must create and maintain a positive relationship with their readers.[49]

How could these statements be wrong? Let's look at how such a narrow conception of knowing how to write for readers perpetuates misunderstandings about how communication works.

Behind each statement is a powerful assumption informed by the container and conduit metaphors:

Figure 5.1. Starbucks Coffee Company terms of use. *Source*: Starbucks, accessed September 8, 2023, https://www.starbucks.com/terms/starbucks-terms-of-use/.

Last Revised: May 2023

PLEASE READ THESE TERMS OF USE CAREFULLY. ACCESSING OR USING THIS WEBSITE CONSTITUTES ACCEPTANCE OF THESE TERMS OF USE ("TERMS"), AS SUCH MAY BE REVISED BY STARBUCKS FROM TIME TO TIME, AND IS A BINDING AGREEMENT BETWEEN THE USER ("USER" OR "YOU") AND STARBUCKS CORPORATION (D/B/A STARBUCKS COFFEE COMPANY) GOVERNING THE USE OF THE WEBSITE. IF USER DOES NOT AGREE TO THESE TERMS, USER SHOULD NOT ACCESS OR USE THIS WEBSITE. THESE TERMS CONTAIN DISCLAIMERS AND OTHER PROVISIONS THAT LIMIT OUR LIABILITY TO USER.

*1. Readers must understand the meaning of what is written **exactly** as the writer intended.*

Statement 1 is supported by the assumption that the goal of communication is that the writer's intention—or what is in the writer's mind—can be transferred "exactly" and unchanged to the reader. As we know from the conduit metaphor, this can happen only if ideas and thoughts are objects that can be put in word and sentence packages to be sent to readers to take out and put into their own minds. This statement stakes the success of communication on creating an equivalency between the minds of the writer and the reader, as if that were the main goal of communication. As we learned in chapter 3 from Reddy's alternative metaphorical model for communication, the PO box–style message-passing communication machine, and our own extensive experience with how communication doesn't, in fact, work reliably as a simple conduit, achieving such an equivalency is either not possible or impossible to know if it has happened. To suggest that it can be achieved creates the opportunity for writers to guess widely from within their own biases about what readers may want or need and/or limit the real audience for a piece of writing to the community that the writer is already a member of—the most common outcome.

2. Writing must achieve its **goal with** the reader.

3. Writers must create and maintain a positive relationship **with** their readers.[50]

It's actually difficult to know exactly what the textbook authors intended by these two statements. It could mean that *writers* have goals for their readers that they want to achieve via writing, or that *readers* have goals that they want to achieve via a piece of writing, or that writing (as a product of writers) must work "with"—as in collaboratively—the reader to achieve its goals. If the goals are achieved, then a positive relationship is maintained with the reader. A generous reading of the statements suggests that regardless of the statements' multivalence, the assumption is that the goals that the writer has for the reader are the same as the reader's goals and the impulse is benevolent toward the reader. The problem is that this benevolence is not always consistent with the goal for writing, especially in types of technical writing common to government and industry that mediate explicit power imbalances between citizens and governments, or between customers and large service providers, such as cell phone service companies. In addition, in much science and technical writing the writer, or author, remains either anonymous or, beyond a byline, minimally present in the text. A simplistic notion of a binary reader-writer relationship doesn't usefully account for the full contextual complexities of the writing situation.

In addition, we know that the goals of readers and the goals of writers are not always aligned; in fact, they may be in direct conflict in scientific and technical writing, and may be in conflict by design. For example, in scientific and technical report writing, the goals of the writer, and the research or industry organization they work for, may be in conflict with the goals of the reader. A researcher may need to report inconclusive or weak experimental results in such a way that a funding agency will still maintain its funding. An engineer may need to report inconclusive or "fruitless initially"[51] product testing results to a manager or a customer so as not to cast doubt on the viability of the product. Regardless, statements 2 and 3 place the agency—or power—for achieving, and possibly defining the goals for writing, with the writer, not the reader. This power imbalance is common to and normal to much scientific and technical writing, especially when the readers are less scientifically or technically knowledgeable or empowered than the writers. This power imbalance is the product of institutions built on a technological rationality, as discussed in chapter 4, that privileges technical outcomes over human impacts.

Let me offer an example familiar to most readers to illustrate how unrealistic it is to assume that shared goals are a generalizable aim in scientific and technical writing. A writing situation where writer and reader goals are not aligned, and may actually be in conflict, involve the ubiquitous terms and conditions statements that we sign when opting in to use a new phone app, website, or wifi network at the airport or a coffee shop. In this case, the goals of the anonymous writers—or corporate voice—of the terms and conditions document are clearly stacked in favor of the writers. The document states the limitations on the reader's use of the technology in such a way that legally protects the provider of the service from any consequences incurred by the reader's use outside of the terms. The reader, who likely has a fairly specific, and possibly urgent, task to accomplish, such as checking their work email at the airport, does not have an option to negotiate these terms and conditions of use. Whether or not the reader can use the service simply depends on whether they "accept" the terms and conditions, even if they limit their ability to accomplish their goals. In fact, it doesn't matter much what the details of the terms and conditions are, since the reader is unlikely to look at them in detail and will just take on the risk of noncompliance that the service provider is placing on them and hope for the best. Often the bind experienced by the user is explicitly acknowledged by the terms and conditions document, even in all caps:

As a reader, confronting a document that shouts at me in all-cap phrases such as "constitutes acceptance," "binding agreement," and "limit our liability to the user" explicitly does not create a good relationship with me—this document is purposefully and intentionally one-sided. A terms and conditions document definitely establishes a complex relationship between the writer and the reader, but it is certainly not one that assumes shared goals and interests.

The terms and conditions example is useful because it is general to most people's experience—most of the time the stakes are pretty low for just clicking "accept." However, for people in vulnerable situations, such as immigrants at the US border with Mexico, much more can be at stake when interacting with government documentation. For example, an initially innocuous-sounding informational handout called "Next Steps for Families" might appear as if its purpose were to support immigrant families navigating entering the United States. However, when we learn that this document was created in the context of the government's 2018 Zero Tolerance Policy that justified separating parents and children at the border so that the parents could be taken into custody, this informational handout is suddenly cast in a very different light. The first two lines of the "first step" make clear that the reader is in criminal custody of a foreign government:

> You are currently in the custody of the US Department of Homeland Security (DHS) Customs and Border Protection (CBP).
> You have been charged with the crime of illegal entry in the United States.[52]

It's hard to imagine a more disempowered, traumatized reader: a likely poor immigrant on the US border, separated from their children, and with recourse to few resources. In this context, the informational document that they have been handed by the US government is purposefully, explicitly, not on their side. Writing this document to maintain a positive relationship with the reader is impossible given the structurally opposed positions that the writer and the reader occupy.

The "Next Steps for Families" document exemplifies a general ethical problem in technical writing practice: Doing audience research in order to write more effective documents for readers is not always practical, and, more to the point, not ethical, when documents are, by policy and design, opposed to the the readers' interests and well-being. I like to point out to writing students shocked by this document that somebody actually had to write it. Let's imagine it was written by a technical writer working for the Department of Homeland Security, likely very far away from where the information sheet was actually handed out to immigrants in detention. Given this, how should the writer approach thinking about the reader? Whose interests are this writer obliged to represent, and in what terms? What do they even know about how the document will be used? What source materials were they given to work with? How much responsibility do they accept for writing a document about a policy that causes harm? Is there any meaning in striving to "maintain a positive relationship with readers" in this scenario?

It might be easy to brush the "Next Steps for Families" document aside as an extreme case of how a simplistic and benevolent notion of writing for readers can have so little utility and cause so much harm. As a tool of oppression, this document functions as a technology of disenfranchisement[53] because its content and design are complicit in the marginalization of a group of people, in this case extremely vulnerable people seeking asylum at the US border. However, rather than rest easy in the notion that most of the scientific or technical writing that is common to the spheres of research or industry wouldn't have the potential to cause this much harm, I suggest that we should embrace what this example is teaching us: that grappling with power imbalances between readers and writers is exactly what all scientific and technical writers do most of the time. To grapple honestly requires a very deep understanding of the rhetorical situation that the document serves—well beyond simply thinking about who the reader is in superficial terms: age, demographics, first language, nationality, race, and ethnicity. Understanding the position, needs, and desires of a reader also requires understanding the systemic structures that have brought the need for the document about in the first place: the policies, the politics, and the history. In addition, it requires considering the systemic structures that have positioned the reader in the situation where they will encounter the document and the other players in the situation beyond the reader. Who else will be there reading the "Next Steps for Families" document with the immigrants and influencing its interpretation and use—guards, agents, social workers, other immigrants? In addition, what will happen to the reader both before and after the document is encountered? Finally, what is the actual physical environment that it will be experienced in, such as low light, extreme cold or heat, or dampness? All of these questions call for a much deeper understanding of the rhetorical situation and a writing response that requires more than adherence to conventional style guidelines. What if we understood the high stakes of doing harm that is so obvious in this case as a potential in all of our writing? How might that outlook challenge the Communication Metaphor?

However, even assuming a generous reading of *Writing Science Right*'s advice for taking into account the reader's goals and interests that includes the broader situation, a new problem shows up when we look at how this advice is meant to be achieved. Here are the headings from the textbook's chapter 2, "Achieving a Readable Style: Learn Techniques for Clear, Concise, Active Writing":

1. Write Squeaky Clean Prose.

2. Use Active Voice for Clarity."[54]

In chapter 2 we find that the openness of the rhetorical impulse to be reader-centric is closed down again by very tightly regulated rules for writing style. The title of the chapter links a reader-centered approach to the windowpane theory of language (clear language is invisible language) and conventional stylistic advice to write concisely and to use active verbs. These instructions to students reinforce that in communication it is the packaging of thoughts or ideas that matter most, that is, the container metaphor for language. The container metaphor assumes that when expressed in spoken or written language, thoughts and ideas are independent of their packaging. In addition, the conduit metaphor assumes that once packaged, thoughts and ideas will be transferred seamlessly and unchanged into the mind of the receiver, or the audience of the message. Therefore, in a container-and-conduit-metaphor–saturated language system, such as English, requests to improve the packaging of thought to improve clarity are considered to be at least benign, and, at best, benevolent, toward the interests of the person or group striving to communicate their thoughts and ideas and toward the audience's receiving them.

We now know via the Standard English fairy tale that achieving clarity in writing is a matter of polishing language to conform to a conventional style for readers or listeners that conforms to Academic English, Standard Written English, or White Mainstream English (WME).[55] But when readers or listeners aren't fluent in SWE or WME, then what might initially appear to range from being at least benign, and at best beneficial, advice turns into a purposeful effort to exclude. In other words, conventional notions of what it takes stylistically to achieve clarity are far from universally held. To assume that they are, as Greenfield would agree, is a form of racist thinking.

Writing Science Right is just one example of a conventional technical and professional writing textbook that professes a reader-focused—or rhetorically aware—approach to teaching writing. Like many others, it confuses a sincere desire to teach students to relate to their audiences empathetically with a full accounting of how social, ideological, and power contexts shape knowledge and language use. That is, learning the conventions for what makes for clear or correct writing for a particular audience is not enough on its own to ensure openness when those very conventions are ideologically informed to privilege one type of knowledge and one type of language use over another. Students who learn how to be savvy about putting their messages in word and document containers to be sent over to a variety of audiences that may vary in level of education, age, context of use, or outlook toward technology do not actually operate outside of the container and conduit conceptual metaphors or the windowpane theory of language. Instead, students of writing are simply learning how to work *within* the Communication Metaphor in a more nuanced way.

Textbooks: Teaching Expediency for a Technological Society

How the Communication Metaphor is perpetuated via textbooks goes beyond the reiteration of the container and conduit metaphors—we also have to consider how technical writing textbooks teach, often uncritically, the values of expediency and productivity as the justifications for learning how to write and as the fundamental shapers of technical style. In a sense, textbooks for scientists, engineers, and technical professionals explicitly celebrate language use that perpetuates "the ethos of a technological society."[56] In other words, what we learned in chapter 4 about technical writing as the currency of a technological rationality that, by design, sidelines human concerns is coded into conventional instruction in the personal and organizational benefits of learning technical writing.

Whether they need to or not, most technical writing textbooks open with a justification for why students should spend the time to develop their written communication skills. Justifications for learning how to write well for scientists, engineers, and other technical professionals are often outcome oriented and focused on the benefits to the individual, the organization, or the profession. For example, textbooks promise students:

> Only by writing (and publishing) can you advance in your career—get cited, get fellowships, get hired, get promoted.[57]

> Surveys over the past three or four decades have shown consistently that employers want to hire people who can communicate.[58]

> The facts of life in the working world are simple: the better you communicate, the more valuable you are.[59]

> Would companies rather not have to spend that money [on teaching employees how to write]? Yes.[60]

Justifications for learning to communicate vary little and tend to be brief because it is assumed that no great effort is necessary to justify an endeavor that appears to be so self-evident and commonsensical, at least to the writers of the textbook. They tend to follow an underlying catechism for why it is important to study and learn technical communication: Effective work=effective communication=happy (profitable) organizations=worker promotion=more money for all.[61] In this catechism we can see the tight weaving of the logic of expediency with the purpose of justifying the study of technical communication. In chapter 4 the origins of this logic were traced to Bacon and Locke and their desire for a more outcome-focused scientific enterprise that benefited society and to the economic and technological revolutions of the Industrial Revolution. We saw how *expediency*, as the logical justification, or ideological warrant (i.e., widely agreed on supporting reason), yokes rational thought and the style of scientific and technical language to *morally* justified aims of productivity and efficiency. So far, the story is fairly straightforward, except for the small matter of ethics and the big lie about language at the heart of it.

Let's start with the small matter of ethics. Aristotle understood language and rhetoric to be embedded in and formed by the messy, contentious contingencies of human action rather than dictated by the rational rules of logic, which he saw as a completely different human faculty. Aristotle understood rhetoric as a *praxis*, or social action, that is shaped by *phronesis*, or the ability of the speaker or writer of rhetoric to reason about what is good for the community. On this point Aristotle and the catechism agree: What is good for the community is defined collectively, and good writers will write within those bounds. However, Aristotle also argues that rhetoric, unlike logic, is an activity that centers and prioritizes human interests over abstract consistency. This means that what is good for the community must always and already include human concerns as well. However, technological rationality has required that we evolve a style of writing for industry, government, and research that suppresses the messy, human aspect of language use so that language doesn't get in the way of productivity and efficiency, which are both beneficial outcomes for organizations. As I've said before, technical writing's greatest superpower is our belief that it is, *by nature,* "short and to the point, with the facts only, no opinions." Such a belief relieves individual writers of the moral burden of the exclusion of human concerns from technical rhetoric: Relax, it's actually required. We saw the ultimate consequences of this belief in the van-refitting technical memo in chapter 4.

And the big lie that all technical writing students must confront: that it is actually *possible* to be "honest," "clear," "objective," and "neutral," all at the same time while writing in the best interests—or for the *good* of—your company or research group. The big lie might be a comforting idea to adopt in a technical writing course, but working people know that it is a much harder writing task to execute successfully. We saw this in the example of the engineer struggling with how to represent sketchy technical testing results and his landing on the vague phrase "fruitless initially" to accommodate the concerns of both the managers and the customer. This engineer's personal struggle with the lack of an acceptable neutral and objective fact to report shows how savvy technical writers have to overcome their training in technical writing to write well. Contrary to what a style handbook might suggest, the best technical writers know how to navigate the "subtle boundary"[62] between the stylistic demands of technical writing and the social and political environment—or the *good* of the community—that they are writing in service of. It's actually really messy out there. If you are an engineering manager, you probably aim to hire engineers who are savvy, and realistic, about navigating this boundary.

A post–Communication Metaphor world, however, would delete even the "subtle boundary" between an idealized technical style and the realities of writing in the real world. Is it possible to live in a world shaped by the ideology of expediency and technological rationality and at the same time unwind the big lie that technical writing can be "honest," "clear," "objective," and "neutral" at the same time? The answer is yes, and it involves elevating the agency of the technical writer and foregrounding how much they have the power to shape knowledge via communication. When linguists write technical writing textbooks, they get it:

> Informative documents are generally thought of as factual and objective. . . . But events can be recorded in different ways and from different points of view. . . . As the writer . . . you select which details you include and in which contexts you present them. . . . Informative documents derive authority from the fact that they *seem* logical, as if any competent professional would have reached the same conclusions or made the same observations.[63]

Thankfully, technical writers are more than trained robots that just recycle the accepted uses of language and argumentation determined by their company, field of research, or technical writing handbook; they are living, breathing, motive-driven, empathetic humans with the ability to make careful, subtle language choices that are responsive to complex rhetorical situations.

Given That Linguists Don't Run the World, We Have CLA

Linguists, of course, don't run the world, even if they write good technical writing textbooks. As a result, the Communication Metaphor persists in all of its elements, including the Standard English fairy tale; the conduit, container, windowpane, and transmission metaphors; the ideology of expediency; the black boxes of "communication" and "clarity"; and a focus on correctness—and it likely will for some time. So, if this is the case, we have to consider: What should scientists and technical professionals be learning in the writing classroom?

A very short answer to this very large question is critical language awareness (CLA). While this is not a book about how to teach technical writing, it's important for readers to know that the people who think very carefully about how to teach scientists and technical professionals how to write also think very carefully about what is the right thing to do by them, and, by doing this thinking, have come up with imperfect but practical-for-now solutions.

In a nutshell, critical language awareness is an approach to teaching writing that aims to productively navigate the tension between pragmatism (i.e., what students need to write today) and progressivism (i.e., how the world will write in a more just tomorrow).[64] It navigates this tension by teaching students explicitly about how language works linguistically, including the differences between a prescriptive (teaching the "right" way to write) and a descriptive (teaching how language is actually used) approach to studying language. Other topic areas include the relationship between language variety and standardized languages and the implications of this relationship for power dynamics and social justice. It's basically about letting students in on how the language puppet show works[65] so that they can join the show and speak and write from an informed and empowered position.

As an approach, CLA is not new but was born of the turmoil in the UK educational system in the 1960s and 1970s. Due to rapid changes in racial and socioeconomic integration in British society that challenged the traditional English language curriculum in schools, educators and politicians recognized the need for a language curriculum that was more inclusive culturally and linguistically. While the original CLA curriculum was not initially widely adopted in the United Kingdom, largely because it was ahead of its time in terms of what society would accept as a divergence from the more prescriptivist traditional language education, the CLA approach persisted over time and has more recently been embraced and developed at the college level in the United States, especially for teaching first-year composition at universities. Critical language awareness is still being developed as an approach to teaching technical writing, and one aim of this book is to provide a resource to accelerate that process.

Critical language awareness integrates well with the overall aim of this book, which is to make the Communication Metaphor and the harm that it does visible, and to imagine, and ultimately work toward, a post–Communication Metaphor future. In fact, this book would serve students well in a CLA curriculum. Of course, as with all issues of teaching language and writing, there is considerable academic and public controversy about what is the right thing to do—I do not in any way want to diminish the complexity of this issue. To quickly summarize a few of the issues, I've adopted a set of hypothetical questions at the heart of the pedagogical conundrum[66] that a well-intentioned technical writing teacher with full knowledge of the Communication Metaphor might have about CLA and offered answers that speak to an audience outside of the writing teacher and scholar community:

QUESTION: If [the Communication Metaphor] perpetuates historical systems of oppression, then should I avoid teaching it altogether?

ANSWER: While some writing teachers and scholars, especially those *outside* of the field of technical and professional writing, would say yes, the consensus is generally for taking the more middle road reflected in a CLA-informed curriculum.

QUESTION: If I do not focus on academic/standardized language in my writing classes, then what would I be teaching instead? Is my main goal simply to validate what students already know and do as language users?

ANSWER: Certainly throwing the baby out with the bathwater is not the solution for teaching technical writing. A helpful way to reframe the issue of curriculum is to move away from deciding which topics are either in or out and refocus the goal as teaching students to be empowered writers who can wield an informed agency to make good, just writing choices no matter what kind of writing situation they find themselves in.

QUESTION: And if [not perpetuating the Communication Metaphor] is my goal, how do I justify it to my students, who have been told that my class is designed to help them write for academic or professional purposes? Was that false advertising?

ANSWER: The goal of bringing students honestly into the problem of the Communication Metaphor and preparing them to thoughtfully navigate it throughout their professional careers generally supports their expectations even if the class is not structured quite as they expected.

QUESTION: Also, how does a shift away from academic/standardized language *inside* my classroom actually change things in the world, including the academic world, *outside*?

ANSWER: In any large sense, it doesn't, although that's not to say that what happens in writing classrooms (including the textbooks assigned) doesn't matter—changes in classrooms are necessary but far from being sufficient for broad change. As I've argued throughout this book, the Communication Metaphor is not just an artifact of technical writing classrooms in universities—its sources are multiple: linguistic, historical, cultural, economic, and political. While the technical writing classroom is *one* site where the Communication Metaphor is often uncritically perpetuated, common teaching approaches and standard textbooks are symptoms rather than the source of the problem.

QUESTION: Maybe I'm in the wrong profession! I really want to do what is best for students—particularly for those from less privileged backgrounds. But I'm not even sure what that is anymore!

ANSWER: Teaching writing of any kind is a tough profession because language use by humans is *always* personally and politically complicated. The fact that we have *ever* believed that teaching technical writing is not complicated is an artifact of the totalitarian hold the Communication Metaphor has had on us.

So, if CLA is a practical-for-now approach to managing a world ruled by the Communication Metaphor, what kinds of assignments might a technical writing student actually experience in a course? In the briefest terms possible, here are a few sketches of assignments:

1. Collect samples of writing from across the various spheres of your life (personal, work, school, family) and study the differences between them. Think about why they are different and why those differences matter and to whom. Write short style guides for each sphere of writing to document the differences for yourself and others.

2. Collect samples of writing from the professional field or discipline that you are training for and analyze the stylistic and linguistic norms of each document and how they vary across types of writing (genres). Think about how those norms came about and who they benefit or exclude. Talk to an experienced person in that field to get their perspective. Write a short report about your findings according to the stylistic and linguistic norms of your field.

3. Take a scientific journal article, engineering report, or other technical document and rewrite a portion of it outside of the Communication Metaphor. Reflect on what you learned about language and writing from this exercise.

Coda to Chapter 5: Generative AI: What Does It Mean for the Standard English Fairy Tale?

Since I began conceptualizing and researching this book in the summer of 2020, generative AI has evolved from a peripheral issue of interest to some technical communication researchers and faculty as a future-looking technology that has interesting implications for industry to a clear and present danger to academia and the world we know it—at least some people would think so. At any rate, for anyone in higher education, research, or any other knowledge-based profession (e.g., law, medicine), it has very quickly become imperative to take a position on how generative AI will impact, and ultimately shape, the future.

My position at this point is both practical and patient. Like many of my faculty colleagues in technical communication and writing studies who teach university-level writing to prepare scientists, engineers, and technical writers for their careers, I accept generative AI-powered applications, such as ChatGPT, as useful tools that, within limits, can enhance our teaching and the practice of professions in academia and industry. At this moment in 2023 multiple special issues of academic journals and edited book collections are in development as writing researchers and faculty scramble to understand and apply this suddenly ubiquitous technology. These efforts, while an earnest and useful reaction to the suddenness of AI's ascendance to the top of our concerns as teachers, researchers, and professionals, are clearly only the very beginning of the story and will quickly be superseded. It's not our first rodeo either: Looking back to the 1990s, we have to accept how little of today's internet and social media–saturated world it was possible to predict. As such, in the interest of the longevity of the themes in this book, I'm going to limit my discussion of AI to raising a few questions that I don't believe anyone can answer with any authority at this point. Future readers will have more resources to reflect on these questions than I have today:

- Will generative AI enable, and hasten the arrival of, a post–Communication Metaphor world?
- Will generative AI powered text tools double down on the Standard English fairy tale or render it moot?
- Do science, engineering, and technical students still need to learn technical writing?
- If they do (surely, they do!), then what exactly is it important for them to learn?
- What will a technical writing textbook teach? Will handbooks survive at all?
- Will so-called Standard English become over time a convention for how machines write, but not humans? Could AI tools actually liberate humans from the need to learn to write and speak a standard written language?

- Taking machine translation tools into account, will the notion of a standard language be annulled by the capabilities of machines to translate any world language, standard English dialect, or other set of language conventions into anything else?

Chapter 6

Experiments in Imagining a Post–Communication Metaphor World

What would scientific and technical writing look like in a post–Communication Metaphor world? This question is, of course, largely hypothetical. Reversing or undoing the histories, powerful ideologies, cultural and economic systems, cognitive structures of the human mind, and everything else that has been argued for in this book under the banner of the Communication Metaphor is actually not practical—but something else a bit less totalistic might be.

This chapter will offer some experiments in imagining this something else. It begins with a summary of the elements of the Communication Metaphor and then works through three experiments that imperfectly imagine writing in a world free of some of these elements. I encourage you to try the experiments for yourself and also to design new ones that resist elements of the Communication Metaphor not explored by the three experiments. Warning: I found it much harder than I imagined, kind of like asking a fish to get out of the ocean and deal with hauling itself across land. Overall, this chapter is an invitation to imagine and experiment with alternatives.

Summary of the Elements of the Communication Metaphor

Chapters 1 through 5 have argued for the existence, the meaning, the origins, and the persistence of the Communication Metaphor—the tacitly held beliefs and practices about what makes for successful technical writing that stand in as symbolic for a more reality-based understanding of how writing and communication really works. This book has argued for how the Communication Metaphor is a complex of multiple elements with varied origins in history, culture, economics, and more that have merged together into a widely accepted orthodoxy. As a summary, the elements of the Communication Metaphor discussed in this book are:

Chapter 2: Making the Communication Metaphor Visible

- **Black boxing**: Recognizing that *communication* is a word that stands in for the poorly understood complexity of how human beings really share their experiences and knowledge with each other.

- **Nominalization**: As an abstract noun formed from an active verb (to communicate), the word *communication* is a red flag for a common concept that has been displaced from its origins in a shared, concrete situation and widely accepted as meaningful in language use without definition.

- **Style, correctness, expediency, the windowpane theory of language, rhetorical framework or thinking, and conduit metaphor:** Elements of the Communication Metaphor made visible by the coding of statements about what makes technical writing successful:

Chapter 3: What Does the Communication Metaphor Mean?

- **Literary metaphor**: Use of metaphor for explanation and definition in scientific writing and *line* as a literary metaphor for a conduit of communication.

- **The container and conduit cognitive metaphor**s: Ideas are objects that go in containers that travel along "lines" of communication between minds.

- **The windowpane theory of language:** The belief that ideas exist independently of language, so that language is like a pane of glass that one looks through to see the idea. Clear windows—or clear language—allow for the most unimpeded view.

Chapter 4: Where Does the Communication Metaphor Come From?

- **Coin or currency**: Another metaphor for technical writing as the value generator for scientific knowledge.

- **Purity**: Successful technical writing is "purified into neutral invisibility"; it just gets out of the way of the way of scientific or technical truths.

- **Plain Style**: The development of Plain Style by the Royal Society in the seventeenth and eighteenth centuries to serve an increasingly elite and professionalized class of scientific practitioners and to justify and generate wealth for the capitalist and colonial enterprises of the increasingly industrialized West. Promoted the development of a purified scientific language free of the language flourishes and alternative epistemologies (e.g., cosmology, theology, scholasticism) of natural philosophy.

- **Plain language:** A set of stylistic conventions for writing about technical, medical, scientific, bureaucratic, and legal topics accessibly for general audiences, such as consumers, citizens, patients, and clients. In general, favors common terms over jargon. Risks relying on the Standard English fairy tale (see chapter 5).

- **Nominalization in science writing**: Common in scientific writing and promotes the creation of jargon via the conversion of observed scientific processes into concepts that can be abstracted away from the original experimental context. Both essential to the progression of scientific knowledge and one of the sources of increasingly jargon-heavy scientific language.

- **The ethic of expediency**: The yoking of rational thought and scientific and technical language to the morally justified aims of productivity and efficiency.

- **Technological rationality and hyperpragmatism**: Ideologies that narrow the aims of technical writing to promoting efficiency and productivity for readers, rather than promoting higher order ways of thinking and doing, such as reflection or centering human concerns.

- **The transmission model of communication**: A mathematical and technological model of electronic telecommunications that gained cultural hegemony (widespread and uncritical acceptance) as a general model for human communication in the context of America's post–World War II economic and geopolitical rise to prominence. An instance of the conduit metaphor that has been legitimized by technical expertise.

Chapter 5: How Is the Communication Metaphor Perpetuated and Maintained?

- **The Standard English fairy tale**: There is one, most correct way to speak and write English to participate fully in school, industry, government, and public life generally. Teaching Standard English as a lingua franca (common language) in schools and enforcing its use across industry and government has benevolent aims that enable universal access to participating in society.

Experiment 1: Scientists at Work— Working across Shared and Distant Contexts

This section dramatizes[1] the difference in communication between scientists in situations where context is largely shared (working collaboratively in the lab office) and in a situation where much less context is shared (a lab researcher reading a scientific paper). The situation dramatizes how assumptions that scientific authors make about the level of shared scientific training and practical knowledge about science with their readers can leave out critical information that can cause mistakes, dead ends, and delays in scientific research. Stylistic values, such as the idea that scientific writing is "short and to the point, with the facts only, no opinions," powerfully shape what kind of information is put in or left out of a scientific research paper. In addition, the common belief among empirical researchers that "the data speaks for itself" further limits the inclusion of contextual information that may be essential to the understanding of a reader who is not already highly embedded in the context of the research due to being a student, working in a different laboratory, the passage of time, or expertise in a different field. In a post–Communication Metaphor world, minimizing length and maintaining a nonpersonal point of view would matter less than creating a text that more fully represented the context of the scientific work.[2]

The Lab Setting

I (the observer) am sitting in a chair in a smallish basement room with six cubicles and a whiteboard at the front. The faculty researcher, Dr. J, sits in one cubicle with an open laptop, within view of the whiteboard. Two of her students sit apart in separate cubicles, but also with views of the whiteboard. Over the course of the team meeting, the students will move to the whiteboard to write on it and also leave the cubicle room to go across the hallway to the lab to retrieve lab notebooks left on the workbench and to bring back vials of solutions that have already been prepared. The students have been assigned to work out the ratio of Molecule X to the other constituents of a formulation that will eventually be applied to petri dishes of live cells to study the effect of Molecule X on the cells as they age.

As they apply themselves to develop the correct ratio of Molecule X, the students and Dr. J work in silence, mostly in their heads and in consultation with their lab notebooks and a scientific paper of related research that reported recommended ratios for Molecule X. The studious atmosphere is punctuated by sparse conversation:

> DR. J: "We need to calculate the concentration of Molecule X. Do you have somewhere written down the concentration of Molecule X that [our lab colleagues] figured out last June?"
>
> STUDENT 1: Leaves room to go across the hallway to the lab to retrieve a lab notebook where calculations related to the concentrations have been recorded. He returns to the room and starts to flip through the notebook to find the calculations. Occasionally they show the notebook to Student 2.
>
> [five minutes pass]
>
> DR. J: "We need to think about this experiment very carefully . . ."

STUDENT 2: Has been looking through a scientific paper and then approaches the whiteboard. They erase part of the existing writing on the board to make space for new thinking. They start to write ratios of molar weights of the constituents of the formulation that they are working on as informed by the scientific paper.

[five minutes pass]

DR. J: "You already prepared the vesicles with Molecule X, yes?"

STUDENT 2: "15 micromolar of Molecule X—is it good?"

DR. J: "No, it's not good . . . the vesicles are old. The lipids definitely aren't going to be ready."

STUDENT 2: Continues to write ratios of molar weights on the board.

About an hour passes in the lab office in mostly silence between Dr. J and her students because most of the work is going on in the heads of Student 1 and Student 2. This is possible because all three share a significant amount of context regarding the what, why, and how of the scientific experiment. As an observer, who does not share the same scientific context, I am not able to discern much about their decision making or reasoning without creating a significant disruption by asking questions. Since the purpose of this meeting, or my observation of their work, is not to teach me about their science or science in general, I recognize, but do not act to mitigate, my own ignorance.

However, while the room may be silent, significant conversation is going on between Student 2 and the scientific paper being used as a reference tool for the concentration of Molecule X, even though that conversation is not audible to an observer.[3] So that we can listen in, at least imaginatively, on this dialogue, let's take a look at the portion of the scientific paper[4] that served as a resource for Student 2's thinking:

As reported to me later by Dr. J, Student 2 used Table 1 in the experimental methods section of the scientific paper as the reference for their calculations for the proportional concentrations of Molecule X and the other constituents of the formulation:

Experiments in Imagining a Post–Communication Metaphor World / 171

Figure 6.1. Recreated Table 1 with data redacted from the scientific paper used by the students of the proportions of the liposomal components used in the various formulations tested.

Constituent	1	2	3	4	5	6	7	8
Phospholipid A	##.#	##	##	##.#		##.#		
Phospholipid B					##.#		##.#	##.#
C	#	#	#	#	#	#	#	#
D	##	##	##	##.#	##.#			
E						##.#	##.#	
F								##.#
Molecule X	#.#	#	#	#.#	#.#	#.#	#.#	#.#

Table 1 was explained in Section 2.5 in the paper's organizational structure but was supported with only one line of text:

2.5 Molecule X liposomal formulations tested

The different Molecule X liposomal formulations evaluated in this study are described in Table 1.

As a reader of this book, Table 1 and the very brief descriptive statement of Section 2.5 may be adequate for your purposes. As a reader of the scientific paper, your response to Table 1 and Section 2.5 might be "Yes, I see what you did here, keep going with the explanation, I'm on board and I don't have much at stake in what's happening so I'm just going along with it." However, if your purpose as a reader, like Student 2, was to mine the paper for your own experimental design, then there would be a lot more at stake in the relationship between you and the author(s) of this paper. A dialogue between Student 2 and the author(s) might go something like this:

> READER: I see that the different Molecule X liposomal formulations evaluated in this study are described in Table 1. How do I know if I can use these formations for my own experiment?
>
> AUTHOR: We have reported our formulations clearly in Table 1.
>
> READER: But were the experimental conditions of your experiment the same as ours?
>
> AUTHOR: We have clearly reported on what we did in the experimental methods section. It's up to you to determine whether your experimental conditions match ours or not.
>
> READER: But which conditions should I take into consideration? There are eight constituents to the formulation in Table 1, and you declare in the paper abstract that cost and availability of the components were important in your decision making. But cost and availability of the components was not reported in the article—is this additional information I am expected to already know via experience or access to other resources? How do I know how much cost and availability impacted the reported proportions of the components, especially Phospholipids A and B?
>
> AUTHOR: We concluded via results from electron microscopy that we had developed an optimized formulation for Phospholipid B based on our measurements of the toxicity of the formulation on living cells. This was the aim of the experiment.
>
> READER: But how do I know if the optimized formulation will work for my experiment? I haven't heard of Phospholipid B.
>
> AUTHOR: I can't give that reassurance. You'll have to try it.

READER: I already did and it didn't work. We are trying to figure out why.

AUTHOR: I can't help you any further.

PUTTING CONTEXT BACK INTO SCIENTIFIC WRITING: A POST–COMMUNICATION METAPHOR ALTERNATIVE

While going to this paper was a reasonable way for Dr. J's students to start working on finding the correct formulation of concentrations for their own experiment, the ratios of Molecule X reported in the source paper turned out to be wrong for their experiment. When Dr. J's students created the lipid vesicles and viewed them through a fluorescence microscope, the vesicles were not perfect round balls, as they should have been, but instead were malformed and crushed—the Molecule X concentration had been too high. While there was some discussion regarding why the concentrations of constituents as reported in Table 1 didn't work, it was assumed by Dr. J, although not known for sure, that it was due to the fact that they used a different phospholipid constituent (neither A nor B in Table 1) in their experimental work than what was reported in the paper. As a result, the decision was made to redo making the liposomal formulations using a concentration of Molecule X that was not reported in the paper but that Student 1 had had success with in previous experimental work. In addition, Dr. J suggested redoing the work using one of the same phospholipids that had been reported in the paper (Phospholipid A), but not the one that the scientific paper had concluded was optimal (Phospholipid B) for the purpose of creating vesicles. Over these course of events, two weeks had passed.

Here we want to step back and ask if there is actually a problem in this situation, and, if there is, what is the role of the Communication Metaphor in this problem? Is it a problem that Dr. J's lab lost two weeks of time due to the experimental dead end the paper sent her students down? This dead end was probably not a result of the intentional withholding of information or purposeful misinformation by the paper's authors—overall, the scientific paper followed the conventions and norms of credible scientific writing and was published in a prestigious scientific journal. Also, experimental stops and starts and dead ends are a normal part of doing laboratory science, and no one in Dr. J's lab seemed stressed out or outwardly concerned by the lost time. But is it *necessary* to lose these two weeks to a miscommunication born of the omission of information from Section 2.5? Section 2.5 is one statement long—it is a black box that offers nothing to readers about the complex process and messy controversies that shape the neat data in Table 1. Did the authors of the scientific paper need to withhold the information that would have clued Dr. J's students into what they needed to know to avoid the experimental failure? What if the paper's authors were able to anticipate and answer the questions raised in the dialogue with readers?

What if Section 2.5 had included more than one sentence and had explicitly answered the reader's questions in the dialogue with the authors about Table 1? What if the authors of the scientific paper didn't assume that readers would have the right training or share enough information about the experimental context to apply the paper to their own experimental design? What if the paper had relayed the proportions of the components in Table 1 in a manner that did more to connect the reported components to the detailed situation and context of their particular lab and study?

For example, what if Table 1 was supported with text like this:

2.5 Molecule X liposomal formulations tested

Table 1 reports the proportions of the constituents of the Molecule X liposomal formulations evaluated in this study. We purposefully used constituents of variable availability and cost. This matters because this is a study for industry and it needs to be cost-effective to scale. The formulations with the purer, higher cost constituents with slightly different molecular structures may work for a research lab and produce less confusing results for scientists, but the commercial interest will have to redo it with a constituent at a cost level that can scale. For labs aiming to use Table 1 as a source document for proportions of liposomal formulations for their own experiments, know that the constituent proportions have been tested for the common lab Phospholipid A, but found to be optimal for Phospholipid B, a cheaper and more industrially viable phospholipid. In fact, Phospholipid B turned out to be 50 percent more successful than Phospholipid A, which is explained in detail in the results section. A complicating fact for some labs may be that Phospholipid B is not available from the American lipid supplier most commonly used by research labs—it is available from a German company that specializes in selling industrial quantities of phospholipids. Another caution: The proportions of Molecule X in Table 1 cannot be assumed to work for Phospholipid Y, which scientific labs may have on hand because it has been used as a standard constituent of lipid formulations in scientific research. In the opinion of the authors of this paper, for experiments aiming to test proportions of constituents for liposomal formulations for applications related to industrial production, the best use of time and resources is to rely solely on Phospholipid B, even though scientific research labs may prefer Phospholipid A because it is more standard for research experiments and produces cleaner results. Phospholipid B can be obtained from the German supplier by making a special application to obtain nonindustrial quantities, but plan an additional four weeks into the experimental timeline for delivery.

This text, which has 320 words in comparison to the thirteen-word original, includes information that many scientific readers may find superfluous. These readers may exclaim that they already know that in research related to the delivery of drugs to cells there is an inherent industrial application in the pharmaceutical industry and of course studies would be influenced by that. They may also argue that they would assume that the proportions reported in Table 1 are only relevant for the exact constituents listed in the table. But what if readers don't make this assumption? Did Dr. J's students know these aspects of the paper's scientific context? Why did the authors of the paper rely on these assumptions rather than spelling out the whole situation? Is saving words that important? Could an additional fifteen seconds of reading time have saved two weeks of lab time? Could the additional text have made this scientific paper more inclusive of Dr. J's students, and other readers with less knowledge and experience in this area of research?

If we return to the mantra that has rung through the chapters of this book, that technical and scientific writing should be "short and to the point, with the facts only, no opinions," and test the alternative Section 2.5 against it, we find that the alternative passage violates all aspects of the statement: It has more than twenty-four times more words; it makes multiple points—well beyond, "the data speaks for itself"; and includes an explicitly stated opinion—"In the opinion of the authors of this paper . . ." However, the alternative passage does answer the questions posed by the reader in the dialogue with the authors of the paper. By answering the reader's questions, the passage makes fewer assumptions about what readers know and works harder to include a broader range of readers in its audience. The alternative text may have saved Dr. J's student significant time in the lab by diverting them away from the fruitless application of the reported proportions of Molecule X to a formulation using Phospholipid Y. Certainly, the amount of experimental time saved—whether one hour or two weeks—would be far more than the additional fifteen or so seconds required to read the longer passage.

Other alternative texts for describing Table 1 might include other types of information or rhetorical tropes, such as literary metaphor, analogy, or genres, such as warnings:

First alternative for Table 1: saying what is missing and an explanation via analogy

Table 1 presents one perspective of the constituents of the Molecule X liposomal formulations evaluated in this study. This table very narrowly displays only the proportions of the constituents of the Molecule X liposomal formulations. Missing from this table are the costs of the constituents and a rating of how easy or difficult they are to obtain. Unfortunately, although it would have been handy for researchers replicating this study, information about the cost and availability of constituents was not included in this paper due to the variability in this information over time. Scientists doing experimental work for industrial applications will need to acquire this information separately. The lab stockroom manager may already be able to provide information or quickly find it out. It may help to think about the choice this way: the choice between Phospholipids A and B depends on the aim of the research. Like making the choice between buying expensive high omega-3 fatty acid farm fresh eggs from the farmers market or the cheapest industrially produced eggs from the supermarket to cook an omelet, the difference is in the nutritional value and taste of the product rather than the fact of whether an omelet can be made. Both Phospholipids A and B can produce successful results, with different nutritional values, so to speak. The differences are explained in the Results section.

Second alternative for Table 1: a personal warning about replicability and a suggestion

We don't recommend trying to replicate the Table 1 data as we have had reports from colleagues even before the publication of this paper that it is not replicable. We include Table 1 here as a record of the Molecule X liposomal formulations tested in this study, although several formulations have been omitted because they were only partially successful for our purposes, even though they may be more relevant to your work. Although we stand behind the conclusions of this study, we don't recommend using Table 1 as a basis for calculating the proportions of constituents for your own liposomal formulations if your experimental work has industrial applications. You may waste several weeks of study time to conclude that the Phospholipid A or B won't work efficiently enough for your experiment. Instead, we recommend looking at methods beyond liposomes—the literature has shown quite a bit of success with nanoparticles.

Third alternative for Table 1: an analogy and general warning

Experimental work can be frustrating and requires a lot of patience. Like choosing the fastest checkout line at the grocery store, choosing good constituents for a solution makes a difference not necessarily in the outcome but in the quality of the experimental process. Experimental time can be reduced and mental anguish avoided for graduate students and lab techs by making good decisions about when to use data in a paper as a starting point for experimental work. Table 1 may not be a good starting point for certain experimental work requiring liposomal formulations, especially for industrial applications.

Fourth alternative for Table 1: literary metaphor and personal warning

Liposomes are the envelopes of biochemistry. Just as an envelope carries your electrical bill through the mail slot into your home, or used to before digital delivery (now the envelope is an email with a link to an account login where you can view your bill electronically—not quite such a tight metaphor!), so does a liposome carry a disease-fighting drug into tissue cells. There are many possible formulations of liposomes depending on your research budget, time available, and the planned applications of your research. Table 1 displays the formulations that we tested across eight constituents. Each formulation has its pros and cons, so make sure you read the results and conclusions sections properly to best understand why we came to our conclusions. We don't guarantee that the data in Table 1 is replicable in other experimental work, but we stand by our results.

Experiment 2: Dramatizing the Message-Passing Machine: Specialist-Nonspecialist Communication

This section dramatizes a use of the message-passing machine discussed in chapter 3 (see fig. 3.4), assuming that the exchange of text and images is between a specialist in a field and an interested nonspecialist. The limiting characteristic of the message-passing machine is that the senders and receivers of messages have no view or knowledge of each others' worlds. In chapter 3 an exchange to share information about a favorite apple turns into a negotiation about what a fruit is, since there are no guarantees that the sender and receiver share knowledge about apples. In the end, both parties have to accept that they can never know for sure that they share exactly the same knowledge about apples.

The episode below illustrates an exchange of understanding about an aspect of the specialist's research—the definition of a liposome—without the nonspecialist needing to adopt the technical language of the researcher. The introduction of scientific jargon is held off until the final exchange, when the topic of explanation is finally named. The main point of this dramatization is to show a reversal of the usual order of technical definition, which normally begins with a technical term (what is being defined) and then calls on additional, often not defined, technical terms for explanation. Instead, the communicative act of definition begins with a metaphor.

A brief definition of liposomes from a scientific source looks like this:

> Small artificial vesicles of spherical shape that can be created from cholesterol and natural non-toxic phospholipids.[5]

Given how understanding of the main term (*liposomes*) and the supporting terms (*vesicles*, *cholesterol*, and *phospholipids*) are unlikely to be initially shared between the specialist and the nonspecialist, using the message-passing machine to reach mutual understanding of a standard definition would require the passing of many notes and take a very long time. But is the specialized terminology really all that important, at least initially? What if all that mattered was that the two conversants left the exchange happy with their understanding of the topic, even if they still did not share a vocabulary of technical terms? The interaction in this dramatization is based on a model of communication that measures success based on the outcomes of the exchange for each participant rather than on the assumption that successful communication is achieved only when the mind of one participant has been replicated in the mind of the other (the conduit metaphor). This dramatization intentionally maintains the separate contexts of the two conversants until they briefly touch at the very end. It draws on metaphor for explanation but assumes that the metaphor is understood differently by each participant:

INTERESTED PERSON: Hello. What do you study?

SCIENTIST: Let me show you. Here is a picture.

Figure 6.2. Doughnut.

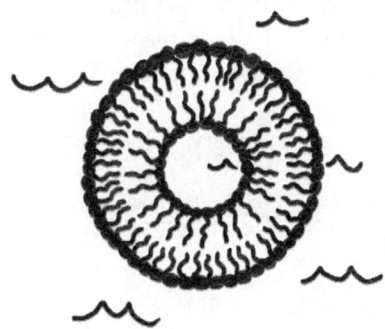

INTERESTED PERSON: Thank you for that lovely picture. I love doughnuts, especially with icing.

SCIENTIST: Yes, I love doughnuts too. This is actually a picture of a jelly doughnut. It has an inner pocket that can be filled.

Figure 6.3. Jelly doughnut in water.

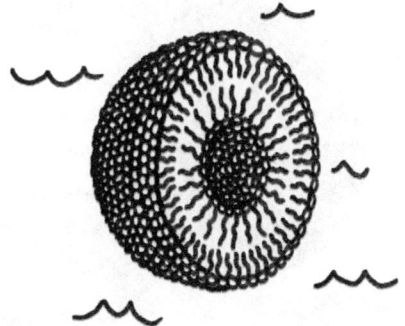

180 / Busting the Myth of the Communication Metaphor

INTERESTED PERSON: Ooh. Tasty. Why is the doughnut floating in water?

SCIENTIST: This is a special kind of doughnut that is made in water.

INTERESTED PERSON: Really? But doesn't the doughnut get soggy? Bleh.

SCIENTIST: No, because the doughnut is made of tiny pieces, or molecules, that both love and hate water. The part that loves water—the head—is on the surface of the donut, and the part that hates water—that tail—is on the inside. The tiny parts align to create a watertight doughnut that never gets soggy.

INTERESTED PERSON: I'm hungry. Where can I get a doughnut?

SCIENTIST: Well, if you were sick then you would get to eat some doughnuts.

INTERESTED PERSON: Really? That sounds like a stomach ache to me and not very healthy. Why would I eat these doughnuts if I were sick?

SCIENTIST: Because these doughnuts can be filled with medicines.

Figure 6.4. Jelly doughnut in water filled with medicine.

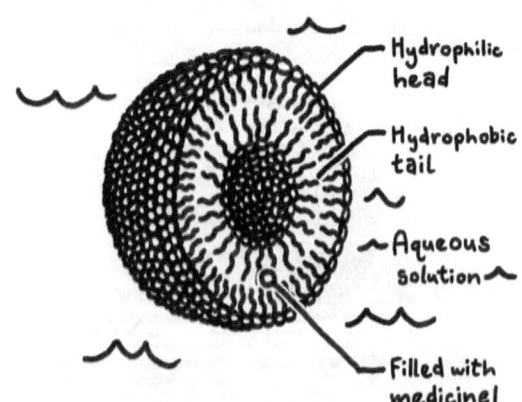

Experiments in Imagining a Post–Communication Metaphor World / 181

INTERESTED PERSON: Ooh. Medicine doesn't usually taste good. Would I still enjoy eating the doughnuts?

SCIENTIST: You probably wouldn't notice the taste.

INTERESTED PERSON: Well, I could probably manage to eat one.

SCIENTIST: One wouldn't be enough. You would probably need to eat millions to get better.

INTERESTED PERSON: Millions? What do you mean? That's a mountain of doughnuts.

SCIENTIST: No, because these doughnuts are very small. They vary in size, but they are measured using nanometers.

INTERESTED PERSON: What is a nanometer?

SCIENTIST: A human hair is 80,000 to 100,000 nanometers wide.

Figure 6.5. Jelly doughnut in water filled with medicine with scale.

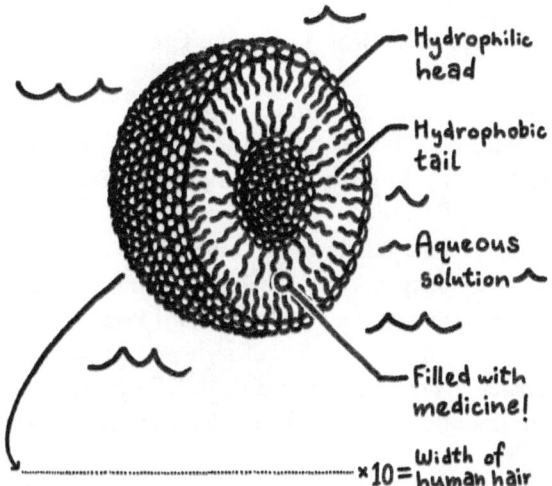

182 / Busting the Myth of the Communication Metaphor

INTERESTED PERSON: Wow, mind blown. Now I can see how I could eat so many doughnuts and not have a stomach ache! Next time I'm sick I'll look forward to doughnuts. What you study is really interesting!

SCIENTIST: Thanks! The doughnuts are actually called liposomes.

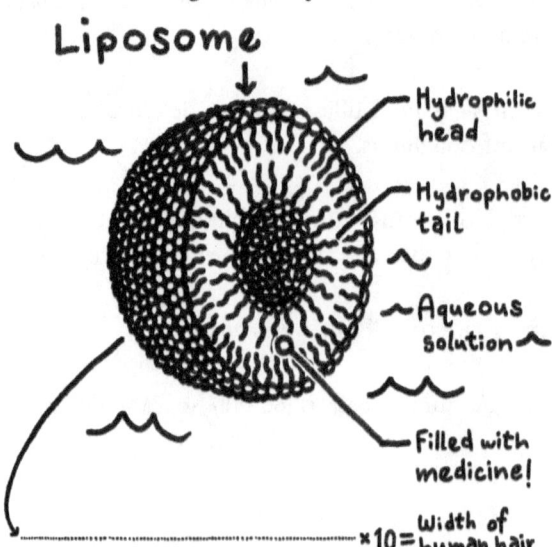

Figure 6.6. Liposome.

INTERESTED PERSON: Right, liposomes. Got it.

Experiment 3: Translating a Scientific Abstract into Nonstandard English Using a Generative AI Tool

In this section I present results of an experiment in which I asked the generative artificial intelligence tool ChatGPT to translate an abstract from a scientific journal into nonstandard Englishes. I created multiple prompts with more or less specificity in terms of style and genre. Chapter 5 called out and demonstrated how the Standard English fairy tale is a major element of the Communication Metaphor. Chapter 5 argued via analysis of common writing handbooks and textbooks that widespread, mostly tacit, belief in the Standard English FairyTale by pretty much anyone who has gone to school, and who is not a linguist, a poet, or another kind of language or writing scholar or professional, perpetuates the Communication Metaphor.

However, imagining, and actually writing samples of, what scientific and technical writing might look like in a world where the Standard English fairy tale has far less power is an incredibly difficult task. It's like asking fish to suddenly live on land–they would have to give up on gills and develop lungs. Since I am also a fish when it comes to scientific and technical writing, but also not a scientist, I found that I could not credibly create scholarly (writing by scientists for scientists) scientific writing not informed by the Standard English fairy tale. So, I turned to a generative artificial intelligence tool to do it for me–not necessarily because it could do it better or more credibly but because it could do it a lot faster and therefore allow my experiment to range farther afield.

The results of the experiment in this section are rife with the traps of linguistic and cultural bias, and I will endeavor to call those out as much as possible. I include this experiment, however, because it raises some very interesting challenges for thinking about what scientific and technical writing might be in a post-Communication Metaphor world, specifically one not built on the Standard English fairy tale. I encourage readers to approach this section as a suggestion to try out a similar experiment yourself rather than accept the results that I received as static or definitive. As we know, generative AI tools based on large language models are rapidly evolving. As you read the responses to the prompts that I inputted into ChatGPT, carefully and critically consider these caveats and limitations:

1. Surely, every result is a caricature of the object of the prompt (e.g., nonstandard English, the style of author Robin Kimmerer, etc.). These results produce just as much fodder for a critique of how large language model–based generative AI tools at this time reproduce and perpetuate bias about standard and nonstandard Englishes[6] as they produce productive illustrations that demonstrate the challenges of generating text outside of the influence of the Communication Metaphor.

2. As you read the texts that ChatGPT produced and the list of notable features of each result, keep in mind that an abstract of a scientific paper is not just an assembly of

scientific terminology, it is also a genre of writing that has a very specific form and function—to summarize the full scientific report so that a reader, a person or a machine, can evaluate the value in reading or acquiring the full text. In general, the results below do not alter the form or the function of the original scientific abstract even as they mix up the words, voice, and sentence construction. In general, you can match the various results sentence by sentence to the original.

3. Most profoundly, and most abstractly, this experiment crashes into the problem of epistemology, or how we know what scientific knowledge is and how we show that knowing via writing. More concretely, the form and function of the abstract of a scientific research report was historically developed to serve experimental science, as we learned in chapter 4. Embedded in the form and function (the genre) of the abstract are elements of the Communication Metaphor, such as the windowpane theory of language (seeing truth "clearly" through invisible language) and nominalization. The topic of the abstract, the development of better liposomes for transporting drugs into cells, is the product of the experimental science that is the outcome of the history and conventions discussed in chapter 4. Would this knowledge even exist without that history? Can, in fact, we write about liposomes at all without the elements of the Communication Metaphor? Can the knowledge gained via experimental science be separated from the tradition of how experimental science has been written about *and* remain equally meaningful?

4. Finally, I ask that scientific readers not discredit this exercise based on the fact that the scientific accuracy of the "translations" is compromised—certainly it is. However, arguing for the need to maintain scientific writing conventions and practices because science can be written only one way or risking being rendered unscientific futher ossifies the Communication Metaphor. The point of this exercise is to imagine an alternative—however imperfect it might appear from our current perspective. The idea that there might even be an alternative is the whole point.

Original Abstract

In this study, layer-by-layer self-assembled liposomes were prepared with different mixing ratios (3:7, 1:1, and 7:3) of sodium alginate (SA) and chitosan (CS), and the morphology, structure, and physicochemical properties of liposomes after layer-by-layer coating were investigated. After coating the liposomes with SA-CS layer by layer (S–C-L), the particle size of liposomes increased from 244.50 to 423.33 nm, the potential changed from negative to positive to 21.20 mV, and the PDI was only 0.393. In the S–C-L 7:3 formulation, the liposomes were uniformly distributed and spherical in size with a clear fingerprint structure. In addition, S–C-L reduced the number of hydrophobic regions within the molecule, allowing the phospholipid molecules to bind tightly into stacks, thus making the liposomes more stable. The S–C-L 7:3 formulation effectively improved the tolerance of liposomes to strong acids and bases, high-salt environments, and UV exposure, and maintained a constant particle size of four hundred nanometers after twenty-eight days of storage. The findings of this study might contribute to the development of polysaccharide layer-by-layer modification of liposomes for a more stable liposomal transport system.[7]

Chat GPT Prompt: Translate This Scientific Abstract into Nonstandard English

In this study, we whipped up some liposome magic layer by layer, mixin' sodium alginate (SA) and chitosan (CS) in different ratios (3:7, 1:1, and 7:3). We dove deep into checkin' out how they look, their structure, and what they're made of after getting coated layer by layer. So, when we coated 'em up with SA-CS layer by layer (S–C-L), the liposomes got bigger from 244.50 to 423.33 nm, switched from negative to positive at 21.20 mV, and kept the PDI low at 0.393.

In the S–C-L 7:3 mix, those liposomes were spread out nice and even, looking round and sporting a clear fingerprint structure. Plus, S–C-L chilled out the hydrophobic spots in the molecules, letting those phospholipids stick together tight, making the liposomes more solid. The S–C-L 7:3 mix did wonders, making these liposomes tough against strong acids, bases, salty spots, and UV rays. Even after twenty-eight days chillin' on the shelf, the particle size stayed a steady 400 nm. Our findings might be the key to leveling up liposomes with some polysaccharide layer-by-layer moves for a supersturdy transport system.

NOTABLE FEATURES OF THIS RESULT:

1. The most interesting feature of this result is the change from the passive voice ("liposomes were prepared") to the third person plural voice ("we whipped up some liposome magic"). In short, the people doing science are put back into the abstract.

2. Words and phrases more commonly associated with spoken vernaculars, such as *mixin'* and "coated 'em up." Perhaps the most significant issue here are the words that the generative AI tool chose as "nonstandard." This result reveals the bias and caricature embedded in the language model. I hesitate to assume who or what is caricatured by these nonstandard language choices, but the choices are laughably, tragically, limited, especially in the choice of words that drop their g (*mixin'*). This result speaks most strongly to how what ChatGPT's language model assumes to be nonstandard English reflects the pervasiveness of the Communication Metaphor.

3. Sentence constructions more commonly associated with spoken vernaculars, such as "So, when we coated 'em up . . ."

4. Everyday terms and phrases that might be considered vague or evaluative, such as "nice and even" and "did wonders."

Chat GPT Prompt: Translate This Scientific Abstract into Nonstandard English Grammar with a Double Negative Construction

. . . The charge ain't negative no more; it's swingin' positive at 21.20 mV, and the PDI ain't no big number, just 0.393.
 . . . Our findings might not just be a maybe; they could be the ticket to pushin' liposomes to the next level with some polysaccharide layer-by-layer moves for a rock-solid transport system.

NOTABLE FEATURES OF THIS RESULT:

1. The double negative constructions in this result do not change the meaning of the original, unless a double negative is read as a net positive value. In practice, the vernacular double-negative constructions in this result are not understood as having a net positive value and so the meaning is not changed.

Experiments in Imagining a Post–Communication Metaphor World / 187

2. As in the nonstandard example above, the choice of words from spoken vernacular, including *ain't* and the dropped *g*'s are evidence of the bias and caricature built into the language model, as if the double negative construction is only a feature of a so-called nonstandard spoken vernacular, which we know from chapter 5 that, historically, it is not.

Chat GPT Prompt: Rewrite This Scientific Abstract in the Style of Scientist and Native Author Robin Kimmerer[8]

. . . This concoction, the S–C-L 7:3 blend, emerged as a resilient guardian, standing tall against the challenges of potent acids, bases, salt-laden realms, and the unrelenting gaze of UV exposure. A steadfast particle size of four hundred nanometers endured, untouched after twenty-eight days of repose on the shelf. The revelations of this study, akin to whispers from the natural world, offer a potential path. They beckon toward a future where polysaccharide layer-by-layer transformations may cradle liposomes in a harmonious dance, creating a vessel of stability for a serene journey through the currents of liposomal transport.

NOTABLE FEATURES OF THIS RESULT:

1. Agency in this result is attributed not to the research team but to the natural phenomena being observed: "This concoction . . . emerged as a resilient guardian," "A steadfast particle size . . . endured . . ."

2. The outcomes of this study are attributed to the natural world speaking to or through the researchers: "They beckon toward a future . . ."

3. While superficially and crudely construed, this result points toward an alternative epistemology for science writing where the source of knowledge and wisdom is the natural world rather than the experimental science conducted by the authors.

Chat GPT Prompt: Rewrite This Scientific Abstract into the Genre of Natural Philosophy, Like Aristotle

. . . Within the alchemical formulation of S–C-L at a ratio of 7:3, an equilibrium akin to the celestial spheres manifests. The liposomes, adopting a celestial uniformity, manifest a spherical visage adorned with a cosmic fingerprint structure. Remarkably, the observed reduction in hydrophobic domains becomes the cosmic choreography, allowing the phospholipid molecules to conjoin in celestial stacks, fortifying the stability of these ethereal liposomes.

The S–C-L 7:3 formulation emerges as a celestial guardian, demonstrating resilience against the cosmic tribulations posed by potent acids, bases, saline realms, and the unyielding gaze of cosmic UV exposure. Astoundingly, the cosmic particle size remains steadfast at four hundred nanometers throughout a celestial epoch of twenty-eight days on the cosmic shelf.

The revelations unfurled in this contemplative journey may, perhaps, be the celestial whispers guiding the refinement of liposomes through the cosmic dance of polysaccharide layer-by-layer modifications, promising a harmonious trajectory for the ethereal transport systems dwelling within the cosmic realms of liposomes.

NOTABLE FEATURES OF THIS RESULT:

1. This result superficially introduces the notion of alchemy and cosmology into the genre of the original abstract reporting on experimental science. This result sets up a direct conflict between the epistemology of the natural philosophy of Aristotle and experimental science. This result points us to the rich field of the philosophy of science.

2. This result, more than the others, pushes at the limits of meaningfully maintaining any meaning from the original. This result forces us to confront that how we think about and understand science is not separate from what we know about the natural world. The modern study of liposomes is the outcome of experimental science, not natural philosophy. Natural philosophy usefully situated phenomena in a broader, integrated, although limited, model of the natural and metaphysical world, in contrast to the primarily reductionist view (natural phenomena are explained by the properties of their components rather than the emergent properties of biological systems)[9] of experimental science. Both views cannot be meaningfully maintained, at least not without imagining a whole new science. I challenge you to try doing just that. To do so is to follow up on the quest to develop an alternative to having to "choose between an antiscientific philosophy and an alienating science."[10]

Conclusion

The purpose of this chapter has been to initiate the project of replacing the way we write and communicate in technical and scientific fields; however, given how difficult this project really is, a post–Communication Metaphor world in which the conventions for technical and scientific writing do not perpetuate injustice remains far off, even if we can see glimpses of hope for its eventual arrival.

What to Do Next: Action

The success of this book rests on whether we, as technical writing fish, have become more aware of the water—the Communication Metaphor—that we swim in. Of course, who we are and where we are coming from will shape the nature of this awareness and how new or uncomfortable it might feel. For readers who have personally experienced, or have experienced via proxy due to membership in a marginalized or underrepresented community, the harms perpetuated by the Communication Metaphor, such as the exclusion of their language practices in academic or technical settings, this book is likely less about creating awareness and more about putting new words to it. For readers whose life experiences have done less to initiate them into the injustices that are baked into our conventional notions of what makes technical writing successful, this book is about creating a new awareness and a language for talking about injustice to continue the project of challenging and ultimately replacing current writing practices. Regardless of your life experience, it is what you do with awareness that matters. Here at the end of chapter 6, we bring to a close the project of creating awareness and initiate a new one—action.

From an individual perspective, the notion of acting to initiate or participate in change that has the power to overcome existing and entrenched injustice can seem overwhelming, risky, pragmatically fruitless, and even hopeless. Very likely, as soon as you put this book down, you will very successfully write an email, technical report, or grant just as you would have before reading about the Communication Metaphor. Fortunately, however, acting individually is not the only, nor the most effective, approach to participating in change. Real change, the kind that can bring us into a more just post–Communication Metaphor world, has to be achieved via ongoing collective, coalitional action. Coalitional action is a framework for social justice work that prioritizes "working collectively to understand oppression and to spur change"[11] and is guided by an overall ethos of humility, especially from allies who have not directly experienced injustice.

This book is an outcome of a kind of coalitional action—it was conceived not in isolation but as a response to a conversation that included me in the moment but that also includes all of the authors cited in this book, my colleagues and editors who have given feedback that has shaped the argument, and readers of this book. In the preface I shared how Jesse, an indigenous graduate of the master's program in professional and technical writing, made clear to me and his graduation oral exam committee that our program falls entirely within and is wholly complicit in the maintenance and propagation of Western Communication practices (his choice to capitalize). Jesse wanted me to understand that these communication practices, or what have been developed in this book as the elements of the Communication Metaphor, have historically caused, and still do presently cause, his tribe harm in the form of government treaties, policies, and other forms of legalese and bureaucratic writing that have driven them off of their land and otherwise restricted their sovereign rights. Jesse made me fully recognize the complicity of our program in perpetuating these same writing practices, even given the fact that the program curriculum already integrates theoretical reading on social justice and technical communication and encourages instructors to engage with community-based learning—reading smart theory and using technical writing for good ends is simply not enough to disrupt the Communication Metaphor.

Upon reflection, I realized that I could use my positionality and power (both key considerations for coalitional action) as the director of the program and an academic trained in technical communication, rhetoric, and writing studies, to respond to Jesse's statements in order to process my own accountability and to offer a response that could contribute to the disruption of Western Communication, or the Communication Metaphor. As a professional in the language arts (as they usefully say in high school), my job is to use words to explain, define, analyze, synthesize, and argue so that what might not be visible in the rush of the daily, commonsense world that we all inhabit, can become an object of critique, and, ultimately, dismantled. Of course, as an academic and teacher there are also practical actions that I can take, such as reform program curriculum and resist conventional teaching practices and resources in my own classroom. Within my own sphere and given my training, these are things that I can do, or what is considered my "margin of maneuverability"[12] in the framework of coalitional action. But it isn't enough, and many others in other spheres with different training than I have will also have to take action—perhaps that is you. That is the question: What can you do in your sphere with your training to reject and replace the Communication Metaphor? What is your margin of maneuverability? The answer is different for everyone, and, likely, a small step is right in front of you.

If you need a suggestion to get started, here is something small and daily that everyone can do: The next time you are annoyed to identify an error in the text of an email, technical report, or grant you might consider whether it might not be an error but nonstandard language use. Do you know the difference? You might ask yourself, given what you know about the Communication Metaphor, what is at stake for you, the company, or the author for "correcting" it, or for leaving it as is. You might have a conversation with colleagues about the complexities of these choices, and, whether they are willing to entertain a post-Communication world or not, the conversation itself is laying the groundwork for an alternative future.

Certainly, not all action has to be at the word or sentence level. Now that technical and scientific writing is largely digital, it matters who has the authority and access to digital platforms. For example, web sites, blogs, content management systems, repositories, and other types of documentation platforms all have systems for setting up administrative access and powers of gatekeeping. Who has access to these systems and in what capacity (editor, administrator, moderator, etc.)? You might ask this question about the platforms that you use and what the concurrent configuration of administrative privileges mean for who can speak and how. Do the current settings for who can post to a blog or comment thread enable constructive, safe discussion and the exchange of ideas, or do they threaten it and shut it down?[13] And who decides what is the case? For example, if an information technology system is highly centralized, do site owners within the system have the power to set edit and post permissions, or must they go through a centralized IT office whose job it is to protect the interests of the institution according to policies that have been made with legal and security considerations in mind? Coalitional action to affect these institutionalized workflows and policies would do even deeper work.

One More Metaphor: Technical Writing as Alchemy

But before the end, a final point of discussion. After sharing the above experiment of using a generative AI tool to translate a scientific abstract into nonstandard English and other styles, I was asked by a knowledgeable writing and language scholar why I didn't just ask the generative AI tool to translate the abstract into plain language—why mess around with the complexities and risks of asking AI to write in specifically *nonstandard* English when it could much more successfully translate it into more common, standard, plain language? This is a great question. My answer to this scholar aims to establish the Communication Metaphor as a concept that exceeds current stylistic and pedagogical approaches to challenging linguistic injustice.

The short answer is that nonstandard writing and plain language are not the same thing and don't have the same aims. Plain language, for all its pragmatic value for making jargon-heavy medical, technical, and legal language more accessible to a general reader, still operates within the Communication Metaphor, as defined by this book. Plain language aims to achieve inclusivity by broadening the audience for technical writing beyond the expert community by reducing or limiting the amount of technical jargon used in texts. This is a good thing, for example, in situations where patients need access to complex medical information in order to make an informed decision. Plain language, however, does not challenge the notion of a standard English or Western ways of making and valuing knowledge.

However, the idea of normalizing what are currently considered nonstandard Englishes, grammars, syntaxes, terminology, and alternative epistemologies (such as indigenous knowledge) in the arenas of science, technology, government, and industry and, ultimately, redrawing the lines around what is considered correct, standard—or plain—is a much more radical, and uncomfortable, proposition. The severe limitations of a generative AI tool in creating a credible artifact that accommodates a common genre of science writing—the article abstract—to nonstandard Englishes, style conventions, and alternative epistemologies underscores exactly how much has to change. Generative AI writing tools are trained, like us, on language use shaped by the Communication Metaphor—it is their water too. If the style, genre, and content of a scientific research article abstract is the outcome of three centuries of the scientific enterprise, we certainly can't just superficially overlay contemporary values for inclusive language and call it a success, and neither can AI. No, the enterprises of science and technology have to fundamentally change, and those changes will ultimately require new, yet to be defined, language conventions to support them—a wholly newly imagined Royal Society Plain Style, if you will.

If new language conventions were defined to support a differently configured scientific and technical enterprise, technical writing would no longer be the coin or currency of science, technology, and industry; but it might be alchemy, if I may introduce a final metaphor. Alchemy, the medieval art and science of transmuting metals into gold, as well as a more generalized process for the spiritual or magical transmutation of feelings, situations, or conflicts into something new and positive, could be a powerful metaphor for technical writing, especially given its multiple meanings and multiple cultural roots. Overall, alchemy construes a notion of transmuting disparate elements into a single new, much more valuable, element. The metaphor of alchemy opens up the question of how the current elements of the Communication Metaphor be transmuted into a more inclusive, more representative of lived experience and more generative approach to technical and scientific language use.

Notes

Preface

1. Scott Richard Lyons, "Rhetorical Sovereignty: What Do American Indians Want from Writing?," *College Composition and Communication* 51, no. 3 (2000): 447–68.

2. Saul Carliner, "Who Technical Communicators Are: A Summary of Demographics, Backgrounds, and Employment," Intercom, January 28, 2019, https://www.stc.org/intercom/2019/01/who-technical-communicators-are-a-summary-of-demographics-backgrounds-and-employment/.

3. Bernadette Longo, *Spurious Coin: A History of Science, Management, and Technical Writing* (SUNY Press, 2000).

4. "CCCC Black Technical and Professional Communication Position Statement with Resource Guide—Conference on College Composition and Communication," Conference on College Composition and Communication, September 2020, https://cccc.ncte.org/cccc/black-technical-professional-communication.

5. Michel Foucault, "The Concern for Truth" (1989), quoted from the frontispiece of Rosina Lippi-Green, *English with an Accent: Language, Ideology, and Discrimination in the United States* (London: Routledge, 1997), a key text in the shaping of my thinking, especially in chapter 5.

Chapter 1

1. The Engineering Criteria 2000 published initially in 1997 by ABET (Accreditation Board for Engineering and Technology) for the accreditation of engineering programs included for the first time an outcome focused on the ability to communicate effectively. The inclusion of communication outcomes had a far-reaching impact on the pedagogy and curriculum of engineering programs, including at the University of Utah college of engineering, where I worked as a writing consultant in the chemical engineering department in the early 2000s.

2. Sarah Read, "15. Making a Thing of Quality Child Care: Latourian Rhetoric Doing Things," *Thinking with Bruno Latour in Rhetoric and Composition* (Southern Illinois University Press, 2015), 256.

3. Sarah Read and Michael E. Papka, "Genre Cycling: The Infrastructural Function of an Operational Assessment Review and Reporting Process at a Federal Scientific Supercomputing User Facilit," *Proceedings of the 32nd ACM International Conference on the Design of Communication CD-ROM*, 2014, 1–8.

4. Norman Fairclough, *Discourse and Social Change* (Polity, 1993).

5. Fairclough, *Discourse and Social Change*, 9–10.

6. Pontus Plavén-Sigray, Granville James Matheson, Björn Christian Schiffler, and William Hedley Thompson, "The Readability of Scientific Texts Is Decreasing over Time," *Elife* 6 (2017): e27725.

7. Hillary C. Shulman, Graham N. Dixon, Olivia M. Bullock, and Daniel Colón Amill, "The Effects of Jargon on Processing Fluency, Self-Perceptions, and Scientific Engagement," *Journal of Language and Social Psychology* 39, no. 5–6 (2020): 13–14.

8. Society of Women Engineers, "Employment of Women in Engineering," tables B24115 and B24116, "Detailed Occupation for the Civilian Employed Population 16 Years and Over," accessed August 2022, https://swe.org/research/2024/employment/. Table data from the US Census Bureau.

9. Society of Women Engineers, "Employment of Women in Engineering," table 505.30, "Data from Digest of Education Statistics, 2020," accessed August 2022, https://swe.org/research/2024/employment/.

10. Boeing, "Celebrating Black History," accessed August 10, 2022, https://www.boeing.com/history/bhm.page. This site has since been removed.

11. Wikipedia, "Purpose," accessed June 25, 2021, https://en.wikipedia.org/wiki/Wikipedia:Purpose.

12. Maja Van der Velden, "Decentering Design: Wikipedia and Indigenous Knowledge," *International Journal of Human-Computer Interaction* 29, no. 4 (2013): 308–16.

13. A. Baker-Bell, B. Williams-Farrier, D. Jackson, L. Johnson, C. Kynard, and T. McMurtry, "This Ain't Another Statement! This Is a DEMAND for Black Linguistic Justice!," Conference on College Composition and Communication, National Council of Teachers of English, 2020, https://cccc.ncte.org/cccc/demand-for-black-linguistic-justice.

14. Geoffrey K. Pullum, "50 Years of Stupid Grammar Advice," *Chronicle of Higher Education* 55, no. 32 (2009): B15.

15. Laura Lisabeth, "Strunk and White and Whiteness," *College Composition & Communication* 73, no. 1 (2021): 80–102.

16. T. McKoy, et al., eds., "Special Issue: Black Technical and Professional Communication," *Technical Communication Quarterly* 31, no. 3 (2022): 221–310.

17. *Resolution on the Students' Right to Their Own Language*, a position statement of the National Council of Teachers of English, published November 30,

1974, at the annual business meeting in New Orleans, Louisiana. https://ncte.org/statement/righttoownlanguage/. The statement was drafted by a committee of the College Composition and Communication arm of the NCTE, which represents the teaching of English in postsecondary settings.

18. Bonnie J. Williams-Farrier. "'Talkin' 'bout Good & Bad' Pedagogies: Code-Switching vs. Comparative Rhetorical Approaches," *College Composition and Communication* 69, no. 2 (2017): 230–59.

19. James Paul Gee, "Literacy, Discourse, and Linguistics: Introduction." *Journal of Education* 171, no. 1 (1989): 5–17.

20. See Miriam Williams and Octavio Pimentel, *Communicating Race, Ethnicity, and Identity in Technical Communication* (Routledge, 2016): 7–20. This collection was on the Black PTW scholars' recommended resource list.

21. April Baker-Bell, *Linguistic Justice: Black Language, Literacy, Identity, and Pedagogy* (Routledge, 2020).

22. Baker-Bell, *Linguistic Justice*.

23. Sarah Read and Michael Michaud, "Hidden in Plain Sight: Findings from a Survey on the Multi-major Professional Writing Course." *Technical Communication Quarterly* 27, no. 3 (2018): 227–48.

24. M. Markel and S. Selber, *Technical Communication*, 13th ed. (Bedford/St. Martins, 2020), 12.

25. Strunk and White, *The Elements of Style*, 4th ed. (Pearson, 2000), 77.

26. Mya Poe, "Re-Framing Race in Teaching Writing across the Curriculum," *Across the Disciplines* 10, no. 3 (January 1, 2013): 1–14, https://doi.org/10.37514/atd-j.2013.10.3.06.

27. Chalice Randazzo, "The Exclusionary Potential of 'Professionalism' in Hiring Situations," *Technical Communication Quarterly* 33 (3) (2024): 270–84.

28. H. Samy Alim, and Geneva Smitherman, *Articulate while Black: Barack Obama, Language, and Race in the US* (Oxford University Press, 2012), 171. Quoted in Baker-Bell, *Linguistic Justice*, 20.

29. British Petroleum, "Diversity Equity and Inclusion at bp," accessed August 10, 2022, https://www.bp.com/en/global/corporate/careers/life-at-bp/diversity-equity-and-inclusion.html,. This website has since been removed.

30. Sarah Read and Michael J. Michaud, "Who Teaches Technical and Professional Communication Service Courses?: Survey Results and Case Studies from a National Study of Instructors from All Carnegie Institutional Types," *Programmatic Perspectives* 10, no. 1 (2018): 77–109.

31. George Lakoff, and Mark Johnson, *Metaphors We Live By* (University of Chicago Press, 2008).

32. *Merriam-Webster Dictionary*, "metaphor," accessed August 12, 2022, https://www.merriam-webster.com/dictionary/metaphor.

33. This term "cell" came around because of Robert Hooke's 1665 publication of *Micrographia*, a book about his observations through a primitive microscope. Hooke named the small compartments that make up plants "cells" because of their

resemblance to the rooms that monks live in at a monastary. James Geary, *I Is an Other: The Secret Life of Metaphor and How It Shapes the Way We See the World* (HarperCollins, 2012), 169.

34. Caleb A. Scharf, "In Defense of Metaphors in Science Writing," *Scientific American Blog Network* (blog), July 9, 2013, https://blogs.scientificamerican.com/life-unbounded/in-defense-of-metaphors-in-science-writing.

35. Rebecca Walton, Kristen R. Moore, and Natasha N. Jones, *Technical Communication after the Social Justice Turn* (Routledge, 2019).

36. Walton, Moore, and Jones, *Technical Communication*, 133–56.

Chapter 2

1. To see the images, go to the British Museum website. Record of beer: clay tablet, the British Museum, asset number: 12175400, https://www.britishmuseum.org/collection/image/121754001; record of barley: clay tablet, the British Museum, asset number: 140853, https://www.britishmuseum.org/collection/object/W_1989-0130-2. Images last accessed December 11, 2024.

2. T. P. Moran, *Introduction to the History of Communication: Evolutions and Revolutions* (Peter Lang, 2010), 89–90.

3. Writing evolved largely independently in several cultures: in ancient Egypt (similar time to Sumer), the Indus Valley (2200 BCE) and China (1300 BCE).

4. I am not the only one to hold this view. Twentieth-century Egyptologist James Henry Breasted argued, "The invention of writing and of a convenient system of records on paper has had a greater influence in uplifting the human race than any other intellectual achievement in the career of man." Quoted in Moran, *Introduction*, 80.

5. USDA Fire Service, accessed July 19, 2021, https://www.fs.fed.us/nwacfire/home/terminology.html. Site no longer accessible.

6. Result from Google Forms survey "Writing and Communication in Science and Industry," survey of science and technology researchers and professionals about writing, PI: Dr. Sarah Read, Portland State University, IRB certified exempt, April 13, 2021.

7. "Black box" is a term developed for actor-network theory and is often attributed to B. Latour, *Science in Action: How to Follow Scientists and Engineers through Society* (Harvard University Press, 1987). Also cited in C. Spinuzzi, "Who killed Rex? Tracing a Message through Three Kinds of Networks." In *Communicative Practices in Workplaces and the Professions: Cultural Perspectives on the Regulation of Discourse and Organizations*, ed. M. Zachry and C. Thralls, 45–66. Routledge, 2007.

8. Cited from Latour, *Science in Action*, 3.

9. Located via a COCA search: Ashley N. McLeod, "Why Websites Work: An Examination of Interdisciplinary Agricultural Center Websites," *Journal of Applied Communications* 102, no. 4 (December 2018): article 12.

10. Located via COCA search: Eric Phetteplace and Jeremy Darrington, "A Hybrid Approach to Discovery Services," *Reference & User Services Quarterly* 53, no. 4 (Summer 2014): 291–95.

11. Located via COCA search: Kathleen Ramsey and Barbara Baethe, "The Keys to Future STEM Careers: Basic Skills, Critical Thinking, and Ethics," *Delta Kappa Gamma Bulletin* 80, no. 1 (Fall 2013): 26–33.

12. Located via a COCA search: Kristin Sayeski, "Putting What We Know into Practice," *ACAD: Teaching Exceptional Children* 47, no. 5 (May 2015): 243–44.

13. McLeod, "Why Websites Work."

14. Sayeski, "Putting What We Know into Practice."

15. M. Markel and S. Selber, *Technical Communication*, 13th ed. (Bedford/St. Martins, 2020), 3.

16. Response from Google Forms survey "Writing and Communication in Science and Industry."

17. G. J. Alred, C. T. Brusaw, and W. E. Oliu, *Handbook of Technical Writing*, 12th ed. (Bedford/St. Martins, 2019), 346.

18. Example from Anne Eisenberg, *Guide to Technical Editing* (Oxford University Press, 1992), 33.

19. Suzanne Eggins, *Introduction to Systemic Functional Linguistics*, 2nd ed. (Continuum, 2004), 3.

20. I. Bello, "Cognitive Implications of Nominalizations in the Advancement of Scientific Discourse," *International Journal of English Studies* 16, no. 2 (2016): 19.

21. Bello, "Cognitive Implications," 5. Bello's historical examples come from I. Moskowich, I. Lareo, G. Camiña Rioboo, and B. Crespo, comps., *A Corpus of English Texts on Astronomy*, CD-ROM (John Benjamins, 2012).

22. Bello, "Cognitive Implications," 6; emphasis mine.

23. Linguist M. A. K. Halliday calls the substitution of one type of word for another—such as a noun for a verb—grammatical metaphor. The substitution of verbs by nouns in scientific writing evolved over the course of its history. Why? Because scientific writing expressed a new type of knowledge that was iterative and progressive and that proceeded from a concrete series of events (an observation or an experiment) to abstract generalizable knowledge (scientific results that over time become generalized into facts). For more detail, see Michael Alexander Kirkwood Halliday and James R. Martin, *Writing Science: Literacy and Discursive Power* (Taylor & Francis, 2003).

24. Bello, "Cognitive Implications."

25. Norman Fairclough, *Discourse and Social Change* (Polity, 1993), 179.

26. Fairclough, *Discourse and Social Change*, 183.

27. Response from Google Forms survey "Writing and Communication in Science and Industry."

28. Paola Ceccarelli, *Ancient Greek Letter Writing: A Cultural History (600 BC–150 BC)* (Oxford University Press, 2013), 27.

29. Malcolm Richardson, "The Earliest Business Letters in English: An Overview," *Journal of Business Communication* 17, no. 3 (1980): 19–31.

30. J. D. Peters, *Speaking into the Air: A History of the Idea of Communication* (University of Chicago Press, 1999), 2.

31. Peters, *Speaking into the Air*, 1.

32. Sarah Read, "How to Build a Supercomputer: US Research Infrastructure and the Documents That Mitigate the Uncertainties of Big Science," *Written Communication* 37, no. 4 (2020): 536–71.

33. Davies, Mark. Corpus of Historical American English (COHA), https://www.english-corpora.org/coha/. Accessed February, 2021.

34. Located via search of Corpus of Historical American English (CHAE): Lee, Henry, Report of a Committee of the Citizens of Boston and Vicinity Opposed to a Further Increase of Duties on Importations, Boston: N. Hale. 1827.

35. Located via search of Corpus of Historical American English (CHAE): Schoolcraft's Journal, *North American Review* (July 1822): 224–50.

36. Located via search of Corpus of Historical American English (CHAE): Hoare, Clement, A practical treatise on the cultivation of the grape vine on open walls, Boston: William D. Ticknor. 1837.

37. Located via search of COCA: Susan MacLeod, "Let's Get Personal," *Today's Parent* 20, no. 8 (September 2003): 99.

38. Response from Google Forms survey "Writing and Communication in Science and Industry."

39. Peters, *Speaking into the Air*, 1.

40. Located via search of Corpus of Historical American English (CHAE): Captain David Porter, "Journal of a Cruise to the Pacific Ocean," *North American Review* (July 1815): 247–75.

41. An accompaniment to Mitchell's map of the world: S. Augustus Mitchell, on Mercator's projection (Hinman and Dutton, 1837).

42. Peters, *Speaking into the Air*, 5.

43. Located via search of Corpus of Contemporary American English: Mike Godwin, *Cyber Rights: Defending Free Speech in the Digital Age*, 1st ed. (Times Books, 1988).

44. Sarah Read, "100% Say Writing Is Important to Their Work, but What Harm Does This Uncontroversial Finding Obscure? Early Results from a Survey of Scientists and Technical Professionals about Writing and Communication," Presented at IEEE Conference on Professional Communication. Limerick, Ireland, July, 17–20, 2022, 21–28.

Chapter 3

1. Caleb A. Scharf, "In Defense of Metaphors in Science Writing," *Scientific American Blog Network* (blog), July 9, 2013, https://blogs.scientificamerican.com/life-unbounded/in-defense-of-metaphors-in-science-writing/.

2. The Editors of Encyclopaedia Britannica, "Black Hole | Definition, Formation, Types, Pictures, & Facts," Encyclopedia Britannica, January 25, 2024, https://www.britannica.com/science/black-hole.

3. James Geary, *I Is an Other: The Secret Life of Metaphor and How It Shapes the Way We See the World* (HarperCollins, 2012), 169.

4. J. S. Weissman, "The Epistemology of Cell Biology," *Molecular Biology of the Cell* 21, no. 22 (2010): 3825, https://www.ncbi.nlm.nih.gov/pmc/articles/PMC2982092/.

5. Christopher Wanjek, "Systems Biology as Defined by NIH: An Intellectual Resource for Integrative Biology," *NIH Catalyst* 19, no. 6 (November–December 2011), https://irp.nih.gov/catalyst/19/6/systems-biology-as-defined-by-nih.

6. Geary, *I Is an Other*, 92.

7. Deirdre N. McCloskey, *The Rhetoric of Economics* (University of Wisconsin Press, 1998), 46.

8. W. Croft and D. A. Cruse, *Cognitive Linguistics* (Cambridge University Press, 2004).

9. Geary, *I Is an Other*, 93.

10. Croft and Cruse, *Cognitive Linguistics*, 196.

11. Croft and Cruse, *Cognitive Linguistics*, 196.

12. Croft and Cruse, *Cognitive Linguistics*, 196.

13. George Lakoff and Mark Johnson, *Metaphors We Live By* (University of Chicago Press, 1980), 32.

14. George Lakoff, "The Contemporary Theory of Metaphor," in *Metaphor and Thought*, ed. Andrew Ortony, 202–51 (Cambridge University Press, 1993), 208.

15. Lakoff and Johnson, *Metaphors We Live By*, 3.

16. Lakoff and Johnson, *Metaphors We Live By*, 10. Attributed to Michael Reddy, "The Conduit Metaphor," *Metaphor and thought* 2 (1979): 285–24.

17. Examples 1–7: Lakoff and Johnson, *Metaphors We Live By*, 11.

18. *Merriam-Webster Dictionary*, s.v. "conduit," accessed October 4, 2021, https://www.merriam-webster.com/dictionary/conduit.

19. Croft and Cruse, *Cognitive Linguistics*, 196.

20. Examples 1–2: Lakoff and Johnson, *Metaphors We Live By*, 10–11.

21. Examples 3–6: Reddy, "Conduit Metaphor," 311–12.

22. Reddy, "Conduit Metaphor," 290.

23. Reddy, "Conduit Metaphor," 295.

24. Reddy, "Conduit Metaphor," 293.

25. Reddy, "Conduit Metaphor," 308.

26. David Kmiec and Bernadette Longo, *The IEEE Guide to Writing in the Engineering and Technical Fields* (John Wiley and Sons, 2017), 109.

27. Major General Fowler, chief of signals and communications of the British Army, after the First World War, quoted from *The Rest Is History* (podcast), episode 216 "Pigeons," July 31, 2022.

28. Located in Corpus of Historical American English (hereafter CHAE): Noah Brooks, "Two War-Time Conventions," *Century* (March 1895): 723–37.

29. Located in CHAE: William Dobein James, 1764–1838, "A Sketch of the Life of Brig. Gen. Francis Marion and a History of his Brigade," Published 1821.

30. Located in CHAE: "Poussin on American Rail-Roads," *North American Review* (April 1837): 435–61.

31. Located in the CHAE: Anna C. Brackett, "The Private School for Girls," (April 1892): 943–60.

32. Carolyn R. Miller, "A Humanistic Rationale for Technical Writing," *College English* 40, no. 6 (1979): 611–12.

33. Examples 1–5: Respondents to research survey about writing and communication in scientific and technical contexts.

34. Examples 1–3: from Sue Neuen and Elizabeth Tebeux, *Writing Science Right: Strategies for Teaching Scientific and Technical Writing* (Routledge, 2018), p. 28, 37, 40.

35. For more information about AS-STE100, visit https://www.asd-ste100.org/index.html. You can request a copy of STE issue 9 at this site.

36. ASD-STE100: *Simplified Technical English*, Aerospace and Defense Industries of Europe, no. 8 (April 30, 2021): page 2-1-R11.

37. ASD-STE100: *Simplified Technical English*, page 2-1-P18.

38. Mike Unwalla, "Increase the Clarity of Text by Using ASD Simplified Technical English (STE)," LinkedIn, 2015, https://www.linkedin.com/pulse/increase-clarity-text-using-asd-simplified-technical-english-unwalla/.

39. ASD-STE100, page 2-0-15.

40. David Ritchie, "Shannon and Weaver: Unravelling the Paradox of Information," *Communication Research* 13, no. 2 (1986): 278.

41. Claude Shannon and Warren Weaver, *The Mathematical Theory of Communication* (University of Illinois Press, 1949), 5.

42. Kmiec and Longo, *IEEE Guide to Writing*, 11. To be fair, this textbook, rather unusually for technical writing texts and manuals, presents the transmission model of communication and its limitations in the context of other models, such as cognitive and behavioral models and social and rhetorical models. The transmission model might be culturally prevalent, but it does not account for all aspects of human communication.

43. Daniel Chandler, "The Transmission Model of Communication," Accessed November 2, 2022, http://visual-memory.co.uk/daniel//Documents/short/trans.html.

44. Shannon and Weaver, *Mathematical Theory of Communication*, 96.

45. Weaver in Shannon and Weaver, *Mathematical Theory of Communication*, 97.

46. A flurry of critical articles in the 1970s and 1980s capped the long-coming end of the "Shannon and Weaver" model as a salient model for communication in communication studies or other areas of language study. For a review, see Joel P. Bowman and Andrew S. Targowski, "Modeling the Communication Process: The Map Is Not the Territory," *Journal of Business Communication* 24, no. 4 (1987): 21–34; and Ritchie, "Shannon and Weaver," 278.

Chapter 4

1. Bernadette Longo, *Spurious Coin: A History of Science, Management, and Technical Writing*. SUNY Press, 2000), xii; emphasis mine.

2. Longo, *Spurious Coin*, xiii.

3. Rickard, T.A., *A Guide to Technical Writing*, 2nd ed. (San Francisco: The Mining and Scientific Press, 1910) Quoted in Longo, *Spurious Coin*, xiii.

4. Rickard, *Guide to Technical Writing*. Quoted in Longo, *Spurious Coin*, xii.

5. Bacon, *Instauration, 1620*. Quoted in Longo, *Spurious Coin*, 38.

6. Longo, *Spurious Coin*, 39.

7. Longo, *Spurious Coin*, 51.

8. John Locke, *Essay* (1693), "Epistle" xvi–xvii, in Longo, *Spurious Coin*, 51.

9. Alexander Ross, *Arcana Microcosmi, or, The Hid Secrets of Man's Body Discovered in an Anatomical Duel between Aristotle and Galen Concerning the Parts Thereof* (Tho Newcomb, 1652; Text Creation Partnership, 2011), 6–7, https://quod.lib.umich.edu/e/eebo/A57647.0001.001/1:4.1?rgn=div2;view=fulltext.

10. Jacob Tootalian, personal communication, March 9, 2023.

11. Excerpt from Robert Hooke's *Micographia*, section "Observ. XVIII. Of the Schematisme or Texture of Cork, and of the Cells and Pores of Some Other Such Frothy Bodies," sourced from *The Project Gutenberg eBook of Micrographia*, November 29, 2023, https://www.gutenberg.org/cache/epub/15491/pg15491-images.html.

12. Excerpt from Hooke's *Micographia*.

13. "Francis Bacon and the Royal Society," Whipple Library, Department of History and Philosophy of Science, University of Cambridge, March 13, 2023, https://www.whipplelib.hps.cam.ac.uk/special/exhibitions-and-displays/exhibitions-archive/universal-harmony/bacon.

14. Quoted in Elizabeth Tebeaux, *The Flowering of a Tradition: Technical Writing in England, 1641–1700* (Taylor & Francis Group, 2014), 54, ProQuest Ebook Central, https://ebookcentral.proquest.com/lib/psu/detail.action?docID=3117834.

15. Excerpts from Joseph Glanville's first (1661) and third edition (1676) the *Vanity of Dogmatizing* from the scholarship of R. F. Jones, "Science and English

Prose Style in the Third Quarter of the Seventeenth Century," *PMLA* 45 (1940): 977–1009. Quoted in Tebeaux, *Flowering of a Tradition*, 54–55.

16. See discussion of plain language writing and its applications for BUROC (bureaucratic, unfamiliar, rights oriented, critical) situations in Russell Willerton, *Plain Language and Ethical Action: A Dialogic Approach to Technical Content in the 21st Century* (Routledge, 2015).

17. Denise Tillery, "The Plain Style in the Seventeenth Century: Gender and the History of Scientific Discourse," *Journal of Technical Writing and Communication* 35, no. 3 (2005): 275.

18. Tillery, "Plain Style," 279.

19. The ideal for scientific writing professed by Joseph Glanville in a 1665 address to the Royal Society. Quoted in Tillery, "Plain Style," 278.

20. Tillery, "Plain Style," 278.

21. Margaret Cavendish, quoted in R. Nate, " 'Plain and Vulgarly Express'd': Margaret Cavendish and the Discourse of the New Science," *Rhetorica* 19, no. 4 (2001): 403–17. Cited in Tillery, "Plain Style," 383.

22. Here I am differentiating institutionalized scientific discourse from the more contemporary practice of nonacademic popular science writing, often written in contemporary plain language, which intentionally aims to communicate science to nonspecialist audiences. From the vantage point of the twenty-first century, it's difficult to imagine a time when the differentiation between specialist and nonspecialist audiences was less distinct (essentially prior to the seventeenth century). Our contemporary commitment to the recognition of the distinction between specialist and nonspecialist audiences is an inheritance of this time period.

23. Tillery, "Plain Style," 274.

24. For a summary of the critique of a dominant masculine rationality and the feminist ways of thinking, knowing, and making knowledge that it excludes, see Elizabeth Anderson, "Feminist Epistemology and Philosophy of Science," *Stanford Encyclopedia of Philosophy* (Spring 2020), ed. Edward N. Zalta, https://plato.stanford.edu/archives/spr2020/entries/feminism-epistemology/.

25. Iria Bello, "Cognitive Implications of Nominalizations in the Advancement of Scientific Discourse," *International Journal of English Studies* 16, no. 2 (2016): 5. Bello's historical examples come from from I. Moskowich, I. Lareo, G. Camiña Rioboo, and B. Crespo, comps., *A Corpus of English Texts on Astronomy*, CD-ROM (John Benjamins, 2012).

26. Iria Bello, "Cognitive Implications of Nominalizations in the Advancement of Scientific Discourse," *International Journal of English Studies* 16, no. 2 (2016): 6. Bello's historical examples come from from I. Moskowich, I. Lareo, G. Camiña Rioboo, and B. Crespo, comps., *A Corpus of English Texts on Astronomy*, CD-ROM (John Benjamins, 2012). This excerpt is cited as Gummere, 1822: 237; emphasis is mine.

27. Suzanne Eggins, *Introduction to Systemic Functional Linguistics*, 2nd ed. (Continuum, 2004).

28. M. A. K. Halliday, "On the Language of Physical Science," in *The Language of Physical Science*, ed. J. Webster, 162–78 (Continuum, 2004).

29. Linguist M. A. K. Halliday understands a "register" as a functional variant of a language (such as English). Unlike a dialect of English, which we assume is a different way of making the same meaning shared across many dialects, a register has unique functional grammar that is purposefully tuned to construe a particular type of knowledge and meaning. The importance of nominalization to expressing scientific knowledge is an example of this. For more detail, see Michael Alexander Kirkwood Halliday and James R. Martin, *Writing Science: Literacy and Discursive Power* (Taylor & Francis, 2003).

30. Norman Fairclough, *Discourse and Social Change* (Polity, 1993), 182–83.

31. Richard Fry, Brian Kennedy, and Carry Funk, "STEM Jobs See Uneven Progress in Increasing Gender, Racial and Ethnic Diversity," Pew Research Center, 2021, accessed Appril 20, 2023, https://www.pewresearch.org/science/2021/04/01/stem-jobs-see-uneven-progress-in-increasing-gender-racial-and-ethnic-diversity/.

32. Sapna Cheryan, Allison Master, and Andrew Meltzoff, "Computer Science and Engineering Need Women: Tech Culture and Stereotypes Dissuade Them," *Scientific American* 328, no. 2 (February 2023): 10.

33. Society of Women Engineers, "Employment of Women in Engineering," accessed August 2022, https://swe.org/research/2024/employment/.

34. Society of Women Engineers, "Employment of Women in Engineering," data from *Digest of Education Statistics*, 2020, table 505.30: accessed August 2022.

35. Cheryan, Master, and Meltzoff, "Computer Science," 10.

36. John Locke, *Second Treatise*, 137 quoted in Longo, *Spurious Coin*, 52; emphasis mine.

37. Longo, *Spurious Coin*, 52.

38. Steven B. Katz, "The Ethic of Expediency: Classical Rhetoric, Technology, and the Holocaust," *College English* 54, no. 3 (1992): 255–75.

39. For a full treatment of English Scientific Register, see Halliday and Martin, *Writing Science*.

40. Bello, "Cognitive Implications," 1–23.

41. C. Bazerman, "Reporting the Experiment the Changing Account of Scientific Doings in the Philosophical Transactions of the Royal Society, 1665–1800," in *Shaping Written Knowledge: The Genre and Activity of the Experimental Article in Science*, vol. 356 (University of Wisconsin Press, 1988), 59.

42. Result from Google Forms survey "Writing and Communication in Science and Industry," survey of science and technology researchers and professionals about writing, PI: Dr. Sarah Read, Portland State University, IRB certified exempt, April 13, 2021.

43. Katz, "Ethic of Expediency," 260–61.

44. Sara Mills, *Discourse* (Routledge, 2004), 13.

45. Cicero, *De Oratore III*, xiv. 55. For a free translation of Cicero, go to https://www.attalus.org/cicero/deoratore3A.html (accessed January 11, 2024), which includes the text translated by J. S. Watson (1860).

46. Excerpt from a Nazi memo quoted in Katz, "Ethic of Expediency," 255–75.

47. Katz, "Ethic of Expediency," 256.

48. Another example is the US Immigration and Customs Enforcement (ICE) information form for families in detention while claiming asylum in the United States. This example is developed in chapter 5.

49. Katz, "Ethic of Expediency," 258.

50. Patrick Moore, "Questioning the Motives of Technical Communication and Rhetoric: Steven Katz's 'Ethic of Expediency,'" *Journal of Technical Writing and Communication* 34, nos. 1–2 (2004): 5–29. It's important to note that Moore's critique of Katz is motivated by the offense he takes that technical rhetoric could be inherently dangerous. Moore's approach to discrediting Katz's argument is to critique his analysis of Aristotle as wrong. In 2006 Katz mounted, in the same journal, a robust defense against Moore's arguments, which are satisfying for my purposes here. In my view, Katz's careful, nuanced argument that connects expediency, technological rationality, and technical and scientific writing style is persuasive, independent of methods of Aristotelian scholarship.

51. Katz, "Ethic of Expediency," 270.

52. Katz, "Ethic of Expediency," 258.

53. Katz, "Ethic of Expediency," 262.

54. Andrew Feenberg, *The Social Life of Reason* (Harvard University Press, 2017), 30.

55. This is outlined by Herbert Marcuse, of the Frankfurt school of Marxist thought, in the first chapter of his major work, *One-Dimensional Man: Studies in the Ideology of Advanced Industrial Society* (Routledge, 1964).

56. Marcuse, *One-Dimensional Man*, 11; emphasis mine.

57. John Leland, "How Loneliness Is Damaging Our Health," *New York Times*. April 20, 2022, https://www.nytimes.com/2022/04/20/nyregion/loneliness-epidemic.html.

58. Susan Well, quoted in Carolyn R. Miller, "What's Practical about Technical Writing," in Bertie E. Fearing and W. Keats (Wendall Keats) Sparrow, eds., *Technical Writing: Theory and Practice* (New York: Modern Language Association of America, 1989), 18.

59. "Ideology" is actually a Marxist term and not unique to language study. See Mills, *Discourse*, 26.

60. J. Blake Scott, Bernadette Longo, and Katherine V. Wills, eds., *Critical Power Tools: Technical Communication and Cultural Studies* (SUNY Press, 2007), 9.

61. Thomas Barker, *Writing Software Documentation: A Task-Oriented Approach*, 2nd ed. (Longman, 2003), xxii; emphasis mine.

62. "Make Firefox Your Default Browser," accessed April 27, 2023, https://support.mozilla.org/en-US/kb/make-firefox-your-default-browser.

63. Barker, *Writing Software Documentation*, 1.

64. Jennifer H. Maher, "The Technical Communicator as Evangelist: Toward Critical and Rhetorical Literacies of Software Documentation." *Journal of Technical Writing and Communication* 41, no. 4 (2011): 377; emphasis mine.

65. Shannon and Weaver, *Mathematical Theory of Communication*, 95.

66. J. D., Peters, *Speaking into the Air: A History of the Idea of Communication* (University of Chicago Press, 1999), 23.

67. Peters, *Speaking into the Air*, 23.

68. Peters, *Speaking into the Air*, 9.

Chapter 5

1. Response from Google Forms survey, "Writing and Communication in Science and Industry," survey of science and technology researchers and professionals about writing, PI: Sarah Read, Portland State University, IRB certified exempt, April 13, 2021.

2. Laura Greenfield, "The 'Standard English' Fairy Tale: A Rhetorical Analysis of Racist Pedagogies and Commonplace Assumptions about Language Diversity," in *Writing Centers and the New Racism: A Call for Sustainable Dialogue and Change*, ed. Laura Greenfield and Karen Rowan, 33–60 (University Press of Colorado, 2011).

3. Rosina Lippi-Green, *English with an Accent: Language, Ideology, and Discrimination in the United States* (London: Routledge, 1997), 10.

4. Lippi-Green, *English with an Accent*, 10.

5. Greenfield, "'Standard English' Fairy Tale," 34.

6. Lippi-Green, *English with an Accent*, 8.

7. Greenfield, "'Standard English' Fairy Tale." Laura. 33–60.

8. Greenfield, "'Standard English' Fairy Tale," 35.

9. Greenfield, "'Standard English' Fairy Tale," 36.

10. For individuals for whom spoken languages are inaccessible, there are alternative visual-manual modes of communication, such as sign languages.

11. April Baker-Bell, *Linguistic Justice: Black Language, Literacy, Identity, and Pedagogy* (Routledge, 2020), 3.

12. Laura Aull and Shawna Shapiro, "FAQs about Language and Linguistics in Writing," WAC Clearinghouse, accessed December 9, 2023, "https://wac.colostate.edu/repository/articles/faqs-about-language-and-linguistics-in-writing/.

13. A useful summary of how linguists understand linguistic equality with references can be found at https://wac.colostate.edu/repository/articles/faqs-about-language-and-linguistics-in-writing/.

14. Lippi-Green, *English with an Accent*, 64.

15. Lippi-Green, *English with an Accent*, 56.

16. Lippi-Green, *English with an Accent*, 58.
17. G. J. Alred, C. T. Brusaw, and W. E. Oliu, *Handbook of Technical Writing*, 12th ed. (Bedford/St. Martins, 2019), 162.
18. Alred, Brusaw, and Oliu, *Handbook of Technical Writing*, ix.
19. Greenfield, "'Standard English' Fairy Tale," 47.
20. Laura Lisabeth, "Strunk and White and Whiteness," *College Composition & Communication* 73, no. 1 (2021): 80–102.
21. Lisabeth, "Strunk and White and Whiteness," 80.
22. Robert J. Connors, "Handbooks: History of a Genre," *Rhetoric Society Quarterly* 13, no. 2 (1983): 87–98.
23. Anne E. Greene, *Writing Science in Plain English* (University of Chicago Press, 2013), 3.
24. In a nutshell, as also discussed in chapter 4, Plain Language is a formal stylistic movement within professional and technical writing to make government, legal, and other patient- and consumer-facing documentation intelligible to everyday citizens. When it comes to federal documentation, it's even the law. While Plain Language is related to plain English, it is motivated primarily by the ethical imperative that critical technical information should be accessible to everyday people, especially in crisis situations. To achieve Plain Language, it would be acceptable to use phrases and terminology considered nonstandard to reach an audience for whom so-called nonstandard English was not familiar. See Russell Willerton, *Plain Language and Ethical Action: A Dialogic Approach to Technical Content in the 21st Century* (Routledge, 2015).
25. Greene, *Writing Science in Plain English*, 1.
26. Greene, *Writing Science in Plain English*, 3.
27. Greene, *Writing Science in Plain English*, 3.
28. Gerald J. Alred, Charles T. Brusaw, and Walter E. Oliu, entry on "clarity," *Handbook of technical writing*, 12th ed. (Macmillan, 2019), 65.
29. Alred, Brusaw, and Oliu, *Handbook of Technical Writing*, 162.
30. Greenfield, "'Standard English' Fairy Tale," 49.
31. Alred, Brusaw, and Oliu, *Handbook of Technical Writing*, 145.
32. Ingrid Tieken-Boon Van Ostade, "Of Social Networks and Linguistic Influence: The Language of Robert Lowth and His Correspondents," *International Journal of English Studies* 5, no. 1 (2005): 135–57.
33. Van Ostade, "Of Social Networks," 148.
34. Van Ostade, "Of Social Networks," 145.
35. Van Osade, "Of Social Networks," 148.
36. Van Ostate, "Of Social Networks," 148.
37. Sarah Read and Michael Michaud, "Hidden in Plain Sight: Findings from a Survey on the Multi-major Professional Writing Course," *Technical Communication Quarterly* 27, no. 3 (2018): 227–48.
38. M. Markel and S. Selber, *Technical Communication,* 13th ed. (Bedford/St. Martins, 2020): 6.

39. Markel and Selber, *Technical Communication*, 47.
40. Markel and Selber, *Technical Communication*, 3.
41. Markel and Selber, *Technical Communication*, 221.
42. Markel and Selber, *Technical Communication*, 12.
43. David Kmiec and Bernadette Longo, *The IEEE Guide to Writing in the Engineering and Technical Fields* (John Wiley & Sons, 2017), 4.
44. Example drawn from Cezar M. Ornatowski, "Between Efficiency and Politics: Rhetoric and Ethics in Technical Writing," *Technical Communication Quarterly* 1, no. 1 (1992): 96.
45. Ornatowski, "Between Efficiency and Politics," 98.
46. Examples 1–2: Markel and Selber, *Technical Communication*, 4, 7.
47. Sue Neuen and Elizabeth Tebeaux, *Writing Science Right: Strategies for Teaching Scientific and Technical Writing* (Routledge. 2018), 1.
48. Neuen and Tebeaux, *Writing Science Right*, 1.
49. Neuen and Tebeaux, *Writing Science Right*, 1.
50. Neuen and Tebeaux, *Writing Science Right*, 1.
51. Ornatowski, "Between Efficiency and Politics," 97.
52. Joseph Bartolotta, "Usability Testing for Oppression," *Communication Design Quarterly Review* 7, no. 3 (2019): 2.
53. Natasha N. Jones and Miriam F. Williams, "Technologies of Disenfranchisement: Literacy Tests and Black Voters in the US from 1890 to 1965," *Technical Communication* 65, no. 4 (2018): 371–86.
54. Examples 1–2: Neuen and Tebeaux, *Writing Science Right*, 37, 40.
55. Baker-Bell, *Linguistic Justice*.
56. S. Michael Halloran, "Eloquence in a Technological Society," *Communication Studies* 29, no. 4 (1978): 221.
57. Stephen B. Heard, *The Scientist's Guide to Writing* (Princeton University Press, 2022), ix.
58. Markel and Selber, *Technical Communication*, 6.
59. Markel and Selber, *Technical Communication*, 7.
60. Markel and Selber, *Technical Communication*, 7.
61. Ornatowski, "Between Efficiency and Politics," 92.
62. Ornatowski, "Between Efficiency and Politics," 98.
63. Kmiec and Longo, *IEEE Guide to Writing*, 86; emphasis mine.
64. Shawna Shapiro, *Cultivating Critical Language Awareness in the Writing Classroom* (Routledge, 2022).
65. The CLA curriculum is closely related to, although it developed independently from, a writing-about-writing (WAW) approach to teaching writing. Writing teachers might be interested in how WAW has been adopted for teaching professional and technical writing: Read and Michaud, "Hidden in Plain Sight," 427–57.
66. Shapiro, *Cultivating Critical Language Awareness*, 7.

Chapter 6

1. The scientific scenario and research study dramatized in this section is loosely based on my own observations, with the permission of Dr. J and her students, of Dr. J's lab at work. It has been fictionalized and redacted so that it can be generalized for wider application and have the maximal rhetorical effect as a demonstration of one aspect of science writing in a post–Communication Metaphor world.

2. Readers may remember the thought experiment from chapter 3 of the mechanical machine for note passing that exposes the fallacy that information about the context of a message can pass automatically from one mind to another mind. The note-passing machine (like passing notes through a post office box) must allow for two-way note passing, so that the two people communicating can iteratively test and negotiate their shared assumptions about what they are talking about until they are both satisfied. Importantly, the end game of the note passing is not exactly the same understanding of the question or issue at hand, but for each note passer to be able to apply their own understanding successfully in their own world.

3. In *Science in Action* Bruno Latour dramatized the dialogue between readers and authors of scientific papers in order to make explicit the necessary persuasive strategies of scientific writing, including extensive citation to show precedence and the presentation of inscriptions (raw and processed data) of natural phenomena. In short, Latour dramatized the scientific paper as a chorus of voices set to persuade the lone reader of the credibility and truth of the scientific argument. Bruno Latour, *How to Follow Scientists and Engineers through Society* (Harvard University Press, 1987).

4. All of the text and data from the scientific paper has been purposefully fictionalized and anonymized to fully disguise the source paper. The scientific scenario sketched here has been purposefully fictionalized and generalized beyond the science of Dr. J's lab in service of foregrounding the demonstration of what a scientific writing could look like in a post–Communication Metaphor world.

5. Abolfazl Akbarzadeh et al., "Liposome: Classification, Preparation, and Applications," *Nanoscale Research Letters* vol. 8, no. 1: 102 (February 2013): doi:10.1186/1556-276X-8-102.

6. Susan D'Agostino, "AI Writing Tools Sometimes 'Correct' Students' Dialects. But Fixing the Tech's Shortcomings May Be Easier Than Addressing Human Language Biases," *Inside Higher Ed*, accessed January 25, 2024, https://www.insidehighered.com/news/tech-innovation/artificial-intelligence/2023/07/10/ai-has-language-diversity-problem.

7. Xiangyun Tan, Yue Liu, Xixi Wu, Mengjie Geng, and Fei Teng, "Layer-by-Layer Self-Assembled Liposomes Prepared Using Sodium Alginate and Chitosan: Insights into Vesicle Characteristics and Physicochemical Stability," *Food Hydrocolloids* 149 (2024): 109606.

8. Robin Kimmerer is a scientist, professor, enrolled member of the Citizen Potawatomi Nation, and bestselling author of *Braiding Sweetgrass: Indigenous*

Wisdom, Scientific Knowledge, and the Teachings of Plants and *Gathering Moss: A Natural and Cultural History of Mosses*. Both books are written for a general reading audience and are demonstrations of what Kimmerer calls "a symbiosis . . . between scientific tools and Indigenous philosophy and wisdom." Kimmerer says, "We live in almost an intellectual monoculture which has rendered traditional knowledge invisible and marginalized that knowledge. . . . Though it is the elder knowledge, that is the most solid, grounded, whole knowledge." Quotes from Tim Peterson, "Robin Wall Kimmerer Explains Indigenous Traditional Knowledge," Grand Canyon Trust, April 18, 2023, https://www.grandcanyontrust.org/blog/robin-wall-kimmerer-explains-indigenous-traditional-knowledge.

9. S. Kesić, "Systems Biology, Emergence, and Antireductionism." *Saudi J Biol Sci.* 23, no. 5 (September 2016): 584–91, doi: 10.1016/j.sjbs.2015.06.015. Epub 2015 Jun 27. PMID: 27579007; PMCID: PMC4992115.

10. Ilya Prigogine and Isabelle Stengers, *Order Out of Chaos: Man's New Dialogue with Nature* (Verso Books, 2018). Quoted in Michael Alexander Kirkwood Halliday and James R. Martin, *Writing Science: Literacy and Discursive Power* (Taylor & Francis, 2003), 21.

11. Rebecca Walton, Kristen Moore, and Natasha Jones, *Technical Communication after the Social Justice Turn: Building Coalitions for Action* (Routledge, 2019), 134.

12. Walton, Moore, and Jones, *Technical Communication*, 143. Coalitional action is the second part of an action plan that follows a process of critical context analysis, which is what this book has done for the Communication Metaphor. The purpose of coalitional action is to "understand the technical communicator's margin of maneuverability in regard to the 4Rs [recognize, reveal, reject, replace], that is, based on their privilege and positionality, how they can wield power to take action" (143).

13. Walton, Moore, and Jones, *Technical Communication*, 149–54. This example is informed by a short case study of coalitional action in which a junior scholar, who is a woman of color whose technical writing students maintain a community-building blog for students from underrepresented groups, lost the authority to approve posting privileges to the blog. The change happened when the English department shifted web hosting platforms and the permission setting became centralized within the authority structure of the department. When the junior faculty member asked to regain the ability to approve posting privileges to the blog in order to maintain the safety of the space for the student community that gathers in it, she was refused with the argument that the new system is both more streamlined (read: expedient) and inclusive (ask: for whom?). Rather than put herself at risk by pushing back against the department and university authority structures alone, the junior faculty member convened a coalition of allies external to the immediate issue of the blog site to develop a collective action plan that could resolve the immediate issue of permission setting for the blog and possibly the larger issue of centralizing IT administration decision making.

References

Akbarzadeh, Abolfazl, et al. "Liposome: Classification, Preparation, and Applications." *Nanoscale Research Letters* vol. 8 no. 1: 102 (February 2013). doi:10.1186/1556-276X-8-102.

Alred, G. J., C. T. Brusaw, and W. E. Oliu. *Handbook of Technical Writing*. 12th ed. Bedford/St. Martins, 2019.

Aull, Laura, and Shawna Shapiro. "FAQs about Language and Linguistics in Writing." WAC Clearinghouse. Accessed December 9th, 2023. https://wac.colostate.edu/repository/articles/faqs-about-language-and-linguistics-in-writing/.

Baker-Bell, A. *Linguistic Justice: Black Language, Literacy, Identity, and Pedagogy*. Routledge, 2020.

Baker-Bell, A., B. Williams-Farrier, D. Jackson, L. Johnson, C. Kynard, and T. McMurtry. "This Ain't Another Statement! This Is a DEMAND for Black Linguistic Justice!" Conference on College Composition and Communication. National Council of Teachers of English, 2020. https://cccc.ncte.org/cccc/demand-for-black-linguistic-justice.

Barker, Thomas. *Writing Software Documentation: A Task-Oriented Approach*. 2nd ed. Longman, 2003.

Bartolotta, Joseph.; "Usability Testing for Oppression." *Communication Design Quarterly Review* 7, no. 3 (2019): 16–29.

Bazerman, C. "Reporting the Experiment the Changing Account of Scientific Doings in the Philosophical Transactions of the Royal Society, 1665–1800." In *Shaping Written Knowledge: The Genre and Activity of the Experimental Article in Science*. 59–79. Vol. 356. University of Wisconsin Press, 1988.

Bello, Iria. "Cognitive Implications of Nominalizations in the Advancement of Scientific Discourse." *International Journal of English Studies* 16, no. 2 (2016): 1–23.

Boeing. "Celebrating Black History." Accessed August 10, 2022. https://www.boeing.com/history/bhm.page.

Bonnie J. Williams-Farrier. " 'Talkin' 'bout Good & Bad' Pedagogies: Code-Switching vs. Comparative Rhetorical Approaches." *College Composition and Communication* 69, no. 2 (2017): 230–59.

Caleb A. Scharf. "In Defense of Metaphors in Science Writing." *Scientific American Blog Network* (blog), July 9, 2013. https://blogs.scientificamerican.com/life-unbounded/in-defense-of-metaphors-in-science-writing/.

Ceccarelli, Paola. *Ancient Greek Letter Writing: A Cultural History (600 BC–150 BC)*. Oxford University Press, 2013.

Chandler, Daniel. "The Transmission Model of Communication." Accessed November 2, 2022. http://visual-memory.co.uk/daniel//Documents/short/trans.html.

Cheryan, Sapna, Allison Master, and Andrew Meltzoff. *Computer Science and Engineering Need Women: Tech Culture and Stereotypes Dissuade Them*. *Scientific American* 328 (2):10 (February 1, 2023): doi: 10.1038/scientificamerican0223-10. PMID: 39017527.

Cicero. *De Oratore III*. Text translated by J. S. Watson (1860). Accessed January 11, 2024. https://www.attalus.org/cicero/deoratore3A.html.

Connors, Robert J. "Handbooks: History of a Genre." *Rhetoric Society Quarterly* 13, no. 2 (1983): 87–98.

Croft, W., and D. A. Cruse. *Cognitive Linguistics*. Cambridge University Press, 2004.

D'Agostino, Susan. "AI Has a Language Diversity Problem. Humans Do, Too: AI Writing Tools Sometimes 'Correct' Students' Dialects. But Fixing the Tech's Shortcomings May Be Easier Than Addressing Human Language Biases." *Inside Higher Ed*. Accessed January 25, 2024. https://www.insidehighered.com/news/tech-innovation/artificial-intelligence/2023/07/10/ai-has-language-diversity-problem.

Davies, Mark. *The Corpus of Contemporary American English*. 2008. www.english-corpora.org/coca/.

Editors of Encyclopaedia Britannica. "Black Hole | Definition, Formation, Types, Pictures, and Facts." *Encyclopedia Britannica*. January 25, 2024. https://www.britannica.com/science/black-hole.

Eggins, Suzanne. *Introduction to Systemic Functional Linguistics*. 2nd ed. Continuum, 2004.

Eisenberg, Anne. *Guide to Technical Editing*. Oxford University Press, 1992.

Fairclough, Norman. *Discourse and Social Change*. Polity, 1993.

Feenberg, A. *Technosystem: The Social Life of Reason*. Harvard University Press, 2017.

"Francis Bacon and the Royal Society." March 13, 2023. Whipple Library, Department of History and Philosophy of Science. University of Cambridge. https://www.whipplelib.hps.cam.ac.uk/special/exhibitions-and-displays/exhibitions-archive/universal-harmony/bacon.

Frith, J. "Technical Standards and a Theory of Writing as Infrastructure." *Written Communication* 37, no. 3 (2020): 401–27.

Fry, Richard, Brian Kennedy, and Carry Funk. "STEM Jobs See Uneven Progress in Increasing Gender, Racial and Ethnic Diversity." Pew Research Center, 2021. Accessed April 20, 2023. https://www.pewresearch.org/science/2021/04/01/stem-jobs-see-uneven-progress-in-increasing-gender-racial-and-ethnic-diversity/.

Geary, James. *I Is an Other: The Secret Life of Metaphor and How It Shapes the Way We See the World.* HarperCollins, 2012.
Gee, James Paul. "Literacy, Discourse, and Linguistics: Introduction." *Journal of Education* 171, no. 1 (1989): 5–17.
Greene, Anne E. *Writing Science in Plain English.* University of Chicago Press, 2013.
Greenfield, Laura. "The 'Standard English' Fairy Tale: A Rhetorical Analysis of Racist Pedagogies and Commonplace Assumptions about Language Diversity." In *Writing Centers and the New Racism: A Call for Sustainable Dialogue and Change,* ed. Laura Greenfield and Karen Rowan, 33–60. University Press of Colorado, 2011.
Halliday, M. A. K. "On the Language of Physical Science." In *The Language of Physical Science,* ed. J. Webster, 162–78. Continuum, 2004.
Halliday, Michael Alexander Kirkwood, and James R. Martin. *Writing Science: Literacy and Discursive Power.* Taylor & Francis, 2003.
Halloran, S. Michael. "Eloquence in a Technological Ssociety" *Communication Studies* 29, no. 4 (1978): 221–27.
Heard, Stephen B. *The Scientist's Guide to Writing.* Princeton University Press, 2022.
Hooke, Robert. *Micographia,* section "Observ. XVIII. Of the Schematisme or Texture of Cork, and of the Cells and Pores of Some Other Such Frothy Bodies." Sourced from the *The Project Gutenberg eBook of Micrographia,* November 29, 2023. https://www.gutenberg.org/cache/epub/15491/pg15491-images.html.
Jones, Natasha N., and Miriam F. Williams. "Technologies of Disenfranchisement: Literacy Tests and Black Voters in the US from 1890 to 1965." *Technical Communication* 65, no. 4 (2018): 371–86.
Katz, Steven B. "The Ethic of Expediency: Classical Rhetoric, Technology, and the Holocaust." *College English* 54, no. 3 (1992): 255–75.
Kesić S. "Systems Biology, Emergence, and Antireductionism." *Saudi J Biol Sci.* 23, no. 5 (September 2016): 584–91. doi: 10.1016/j.sjbs.2015.06.015. Epub 2015 Jun 27. PMID: 27579007; PMCID: PMC4992115.
Kmiec, David, and Bernadette Longo. *The IEEE Guide to Writing in the Engineering and Technical Fields.* John Wiley and Sons, 2017.
Lakoff, G., and M. Johnson. *Metaphors We Live By.* University of Chicago Press, 2008.
Lakoff, George. "The Contemporary Theory of Metaphor." In *Metaphor and Thought,* ed. Andrew Ortony, 202–51. Cambridge University Press, 1993.
Latour, B. *Science in Action: How to Follow Scientists and Engineers through Society.* Harvard University Press, 1987.
Leland, John. "How Loneliness Is Damaging Our Health." *New York Times.* April 20, 2022. https://www.nytimes.com/2022/04/20/nyregion/loneliness-epidemic.html.
Lippi-Green, Rosina. *English with an Accent: Language, Ideology, and Discrimination in the United States.* London: Routledge, 1997.
Lisabeth, Laura. "Strunk and White and Whiteness." *College Composition & Communication* 73, no. 1 (2021): 80–102.

Longo, Bernadette. *Spurious Coin: A History of Science, Management, and Technical Writing*. SUNY Press, 2000.

Lyons, Scott Richard. "Rhetorical Sovereignty: What Do American Indians Want from Writing?" *College Composition and Communication* 51, no. 3 (2000): 447–68.

Maher, Jennifer H. "The Technical Communicator as Evangelist: Toward Critical and Rhetorical Literacies of Software Documentation." *Journal of Technical Writing and Communication* 41, no. 4 (2011): 367–40.

Marcuse, Herbert. *One-Dimensional Man: Studies in the Ideology of Advanced Industrial Society*. Routledge, 1964.

Markel, M., and S. Selber. *Technical Communication*. 13th ed. Bedford/St. Martins, 2020.

McCloskey, Deirdre N. *The Rhetoric of Economics*. University of Wisconsin Press, 1998.

McKoy, T., Temptaous and members of the CCCC Black Technical and Professional Writing Task Force. "CCCC Black Technical and Professional Communication Position Statement with Resource Guide." *Conference on College Composition and Communication*, September 2020. Last accessed December 19, 2024. https://cccc.ncte.org/cccc/black-technical-professional-communication.

McKoy, T., et al., eds. "Special Issue: Black Technical and Professional Communication." *Technical Communication Quarterly* 31, no. 3 (2022): 221–310.

Merriam-Webster Dictionary. s.v. "conduit." Accessed October 4, 2021. https://www.merriam-webster.com/dictionary/conduit.

Merriam-Webster Dictionary. "Metaphor." Accessed August 12, 2022. https://www.merriam-webster.com/dictionary/metaphor.

Miller, Carolyn R. "A Humanistic Rationale for Technical Writing." *College English* 40, no. 6 (1979): 610–17.

Mills, Sara. *Discourse*. Routledge, 2004.

Moore, Patrick. "Questioning the Motives of Technical Communication and Rhetoric: Steven Katz's 'Ethic of Expediency.'" *Journal of Technical Writing and Communication* 34, nos. 1–2 (2004): 5–29.

Moran, T. P. *Introduction to the History of Communication: Evolutions and Revolutions*. Peter Lang, 2010.

Resolution on the Students' Right to Their Own Language. A position statement of the National Council of Teachers of English. Published November 30, 1974, at the annual business meeting in New Orleans, Louisiana. https://ncte.org/statement/righttoownlanguage/.

Neuen, Sue, and Elizabeth Tebeaux, *Writing Science Right: Strategies for Teaching Scientific and Technical Writing*. Routledge, 2018.

Ornatowski, C. M. "Between Efficiency and Politics: Rhetoric and Ethics in Technical Writing." *Technical Communication Quarterly* 1, no. 1 (1992): 91–103.

Peters, J. D. *Speaking into the Air: A History of the Idea of Communication*. University of Chicago Press, 1999.

Peterson, Tim. "Robin Wall Kimmerer Explains Indigenous Traditional Knowledge." Grand Canyon Trust. April 18, 2023, https://www.grandcanyontrust.org/blog/robin-wall-kimmerer-explains-indigenous-traditional-knowledge.

Poe, Mya. "Re-Framing Race in Teaching Writing across the Curriculum." *Across the Disciplines* 10, no. 3 (2013): 1–14.

Plavén-Sigray, Pontus, Granville James Matheson, Björn Christian Schiffler, and William Hedley Thompson. "The Readability of Scientific Texts Is Decreasing over Time." *Elife* 6 (2017): e27725.

Pullum, Geoffrey K. "50 Years of Stupid Grammar Advice." *Chronicle of Higher Education* 55, no. 32 (2009): B15, 2009.

Randazzo, Chalice. "The Exclusionary Potential of 'Professionalism' in Hiring Situations." *Technical Communication Quarterly* 33, no. 3 (2024): 270–84. doi:10.1080/10572252.2024.2340432.

Read, Sarah. "15. Making a Thing of Quality Child Care: Latourian Rhetoric Doing Things." In *Thinking with Bruno Latour in Rhetoric and Composition*. Southern Illinois University Press, 2015.

Read, S. "How to Build a Supercomputer: US Research Infrastructure and the Documents That Mitigate the Uncertainties of Big Science." *Written Communication* 37, no. 4 (2020): 536–71.

Read, S. "The Infrastructural Function: A Relational Theory of Infrastructure for Writing Studies." *Journal of Business and Technical Communication* 33, no. 3 (2019): 233–67.

Read, S. *Survey on Writing and Communication in Science and Industry.* Unpublished raw data. Portland State University, 2022.

Read, S. (2022). "100% Say Writing is Important to Their Work, but What Harm Does This Uncontroversial Finding Obscure? Early Results from a Survey of Scientists and Technical Professionals about Writing and Communication." Presented at IEEE Conference on Professional Communication. Limerick, Ireland, July 17–20, 2022.

Read, Sarah, and Michael E. Papka. "Genre Cycling: The Infrastructural Function of an Operational Assessment Review and Reporting Process at a Federal Scientific Supercomputing User Facility." *Proceedings of the 32nd ACM International Conference on the Design of Communication CD-ROM*, 2014.

Reddy, Michael. "The Conduit Metaphor." In *Metaphor and Thought*, ed. Ortony, Andrew, 284–324. Cambridge University Press, 1979.

Richardson, Malcolm. "The Earliest Business Letters in English: An Overview." *Journal of Business Communication* 17, no. 3 (1980): 19–31.

Ritchie, David. "Shannon and Weaver: Unravelling the Paradox of Information." *Communication Research* 13, no. 2 (1986): 278–98.

Saul Carliner. "Who Technical Communicators Are: A Summary of Demographics, Backgrounds, and Employment." Intercom, January 28, 2019. https://www.

stc.org/intercom/2019/01/who-technical-communicators-are-a-summary-of-demographics-backgrounds-and-employment/.

Scott, J. Blake, Bernadette Longo, and Katherine V. Wills, eds. *Critical Power Tools: Technical Communication and Cultural Studies*. SUNY Press, 2007.

Shannon, Claude, and Warren Weaver. *The Mathematical Theory of Communication*. University of Illinois Press, 1949.

Shapiro, Shawna. *Cultivating Critical Language Awareness in the Writing Classroom*. Routledge, 2022.

Shulman, Hillary C., Graham N. Dixon, Olivia M. Bullock, and Daniel Colón Amill. "The Effects of Jargon on Processing Fluency, Self-Perceptions, and Scientific Engagement." *Journal of Language and Social Psychology* 39, no. 5–6 (2020): 579–97.

Society of Women Engineers. "Employment of Women in Engineering." Accessed August 2022. https://swe.org/research/2024/employment/.

Spinuzzi, C. "Who killed Rex? Tracing a Message through Three Kinds of Networks." In *Communicative Practices in Workplaces and the Professions: Cultural Perspectives on the Regulation of Discourse and Organizations*, ed. M. Zachry and C. Thralls, 45–66. Routledge, 2007.

Starbucks. "Terms of Use: Starbucks Coffee Company." Accessed September 8, 2023. https://www.starbucks.com/terms/starbucks-terms-of-use/.

Strunk, Oliver, and E. B. White. *The Elements of Style*. 4th ed. Pearson, 2000.

Tan, Xiangyun, Yue Liu, Xixi Wu, Mengjie Geng, and Fei Teng. "Layer-by-Layer Self-Assembled Liposomes Prepared Using Sodium Alginate and Chitosan: Insights into Vesicle Characteristics and Physicochemical Stability." *Food Hydrocolloids* 149 (2024): 109606.

Tebeaux, Elizabeth. *The Flowering of a Tradition: Technical Writing in England, 1641–1700*, Taylor & Francis Group, 2014.

Tillery, Denise. "The Plain Style in the Seventeenth Century: Gender and the History of Scientific Discourse." *Journal of Technical Writing and Communication* 35, no. 3 (2005): 273–89.

Tootalian, Jacob. Personal communication. March 9, 2023.

Unwalla, Mike. "Increase the Clarity of Text by Using ASD Simplified Technical English (STE)." LinkedIn, 2015. https://www.linkedin.com/pulse/increase-clarity-text-using-asd-simplified-technical-english-unwalla/.

Van der Velden, Maja. "Decentering Design: Wikipedia and Indigenous Knowledge." *International Journal of Human-Computer Interaction* 29, no. 4 (2013): 308–16.

Van Ostade, Ingrid Tieken-Boon. "Of Social Networks and Linguistic Influence: The Language of Robert Lowth and His Correspondents." *International Journal of English Studies* 5, no. 1 (2005): 135–57.

Walton, Rebecca, Kristen Moore, and Natasha Jones. *Technical Communication after the Social Justice Turn: Building Coalitions for Action*. Routledge, 2019.

Wanjek, Christopher. "Systems Biology as Defined by NIH: An Intellectual Resource for Integrative Biology." *NIH Catalyst* 19, no. 6 (November–December 2011). https://irp.nih.gov/catalyst/19/6/systems-biology-as-defined-by-nih.

Weissman, J.S. "The Epistemology of Cell Biology." *Molecular Biology of the Cell* 21, no. 22 (2010): 3825. https://www.ncbi.nlm.nih.gov/pmc/articles/PMC2982092/.

Wikipedia. "Purpose." Accessed June 25, 2021. https://en.wikipedia.org/wiki/Wikipedia:Purpose.

Willerton, Russell. *Plain Language and Ethical Action: A Dialogic Approach to Technical Content in the Twenty-First Century*. New York: Routledge, 2015. https://doi.org/10.4324/9781315796956.

Williams, Miriam, and Octavio Pimentel. *Communicating Race, Ethnicity, and Identity in Technical Communication*. Routledge, 2016.

Index

AAVE. *See* African American Vernacular English
abstraction. *See also* scientific abstract
   nominalization enabling, 99–104
   repurposing and, 100–102
Academic English (AE), 11, 136
   home languages and, 129
   reader-centered approach and, 155
   Standard English fairy tale and, 131
accuracy, 51
action. *See* coalitional action
AE. *See* Academic English
aerospace industry, 79–80, 81, 147–148
African American Vernacular English (AAVE), 13, 130
   correctness assumptions and, 16–17
   double negatives in, 143
   Standard English fairy tale and, 126
AI. *See* artificial intelligence
alchemy, 188, 191–193
analogies
   brick castle, 101–102
   doughnut, 179–182, *179–182*
   windowpane theory *as clarity*, 21
anecdotes, Communication Metaphor illustrations and
   "Better You Than Me!," 9–15
   clarity importance and definition problem, 20–23

   incorrect writing as unprofessional, 15–18, 19
   polite conversation about profession, 9–10
   tribal college and CEO assignment, 18–20
apple-passing scenario, 69–71
Arcana Microcosmi (Ross), 92, 93
Aristotle, 93, 96–97, 157
   Katz reading of, 112–113, 114, 206n50
   scientific abstract translation into genre of, 187–189
   on speech types, 112–113
artificial intelligence (AI)
   plain language and, 191–192
   scientific abstract translated into nonstandard English with, 182–189
   Standard English fairy tale and generative, 162–164
ASD-STE100, 80
Associated Teachers of Technical Writing, 18
assumptions, *complex* of tacit, 30–33
audience (readers)
   audience awareness, 148–149, 155
   clarity as determined by, 22–23
   coding and, 54
   context experiment and, 168–169

audience (readers) *(continued)*
  "fruitlessly initial" and writing for specific, 148–149, 152
  goals of writers and, 151–152
  metaphor creating common ground for, 25
  Royal Society censure of, 97–99, 104–106
  scientific prose for broader, 98, 204n22
  specialist and nonspecialist, 98
authors (writers), reader's judgment of, 141

bachelor degrees, demographics and, 6
Bacon, Francis, 91–92, 95, 106
  Cavendish and, 98
  coin metaphor and, 117
Bazerman, Charles, 108–109
Bell System Technical Journal, 82
Bell Telephone Labs, 82
Bello, Iria, 43, 100–102
Bible, double negatives in King James, 143
biology, reductionist systems in, 61
black boxes (black boxing)
  clarity statements and, 139
  communication as, 33–38, 54, 141
  controversy avoidance through, 43
  in cybernetics, 34–35, *35*, *86*
  definition by repeating, 38–40
  DNA double helix and, 34–35, 43
  Latour and, 34, 35, 198n7
  linguistic, 32–33
  Standard English fairy tale and, 139
  transmission model and, 86
  usefulness of, 35–36
Black Englishes, 130, 143. *See also* African American Vernacular English
black holes, 59–60
Black Lives Matter, 17

Black professional and technical writing (Black TPC), 12–14
blogs, 191
BP. *See* British Petroleum
brick castle analogy, 101–102
British Petroleum (BP), 17
bureaucratic writing, 45, 190

capitalism, 110, 116, 119
Cavendish, Margaret, 98
CDA. *See* critical discourse analysis
CDM. *See* conduit metaphor
cell
  dead metaphors and, 87
  plant "cells," 25, 60–61, 197n33
CERN, Large Hadron Collider, 46
ChatGPT, 163, 183–184
  translation prompts, 185–189
Chaucer, Geoffrey, 143
Church of England, Lowth and, 143
Cicero, 112
Civil War, lines of communication and, 75
CLA. *See* critical language awareness
clarity
  audience as measure of, 22–23
  coding for survey answer on, 53
  as concept, 21
  correctness and, 22
  handbooks use of term, 138–139, 140
  importance assumption about, 20–23
  justification for purity and, 106–107
  lack of definition, 20–23
  as metaphor, 50
  problem with truism of, 78–79
  seventeenth-century inheritance and, 107–108
  Simplified Technical English and, 79–81

Index / 223

survey responses on, 53, 78, 123
windowpane theory and, 78–79, 139
coalitional action, 27, 189–191
 margin of maneuverability in, 190, 211nn12–13
code-switching, 13–14, 134. *See also* encoding
coding
 black box, 139
 clarity statements, 138–140
 Communication Metaphor, 51–54
 definition of, 51
 statements in light of, 140
cognitive linguistics, 62
cognitive metaphor, 58, 77. *See also* conduit metaphor
coin metaphor, 90–95, 117, 167
 software documentation example of, 119
colonialism, 106–107
commodification, handbooks and, 135–141
communicate, PTW black boxes and, 37–38
communication, 24
 advent of personal realm of, 46
 alternative set of assumptions about, 26–27
 as black box, 33–38, 54, 141
 blindness to term of, 34
 defining by repeating black boxes, 38–40
 evolution of, 46–49
 historical context of, 45–49
 in nineteenth-century texts, 46–47
 as nominalization, 8, 32–33, 40–44
 omitted meaning of, 8–9
 red flags, 32–33, 42, 54
Communication Metaphor. *See also* anecdotes, Communication Metaphor illustrations and;
harm, from Communication Metaphor; post-communication metaphor world; writing and communication; *specific topics*
 assignments outside of, 162
 codes applied to statements and, 51–54
 complex of, 30–33
 elements summary, 165–168
 expediency as key element of, 107–115
 illustrations/anecdotes and, 24
 as invisible and ubiquitous, 29–30
 "metaphor" and, 24–26, 57
 metaphorest fire complex, *31*
 perpetuation of, 169
 power and suppression resulting from, 4
 reason for choosing moniker of, 24–27
 source of, 167–168
 statements showing elements of, 109–110
 White dominance maintaining, 40
 workings of, 49–55
complex
 Communication Metaphor, 30–33
 USDA Fire Service definition of, 31
conceptual metaphor, 61
 conduit type of, 65–77
 container type of, 62–65, *65*, 73, 79, 82
conduit metaphor (CDM), *31*, 53–54, 65, 71, 82
 alternative model to, 70, *70*
 clarity and, 140
 developer of, 67–69
 doughnut analogy and, 179–182, *179–182*
 IEEE Guide on, 147
 inside literary metaphor, 74–77, *76*

conduit metaphor (CDM) *(continued)*
   messages sent through pathways in, 66, 66–67
   statements, 72–73
   technical writing courses and, 145–146
   transmission model and, *84*, 84–85
Conference on College Composition and Communication, 18
container metaphor, 62–65, *65*, 79, 82
   technical writing courses and, 145
   textbooks offering pragmatic alternatives to, 73
contexts
   reinstating, 173–178
   shared and distant, 169–170
conversation
   about profession, 9–10
   written language *vs.* face-to-face, 128–130
corpora, 46
Corpus of Historical American English, 47, 76
correctness, 146
   AAVE and, 16–17
   clarity and, 22
   coding for, 53–54
   equated with WEV or SAE, 16
   Native American students and, 19
   professionalism and, 15–18, 19
   social class and, 144
Crick, Francis, 34–35
critical discourse analysis (CDA), 103, 104
critical language awareness (CLA), 159–162, 209n65
cultural memes, nominalization and, 44
currency of productivity, 107. *See also* coin metaphor
cybernetics, black box in, 34–35, *35*, 86

dead metaphors, 61
demographics, employment statistics and, 6, 104–105
Department of Homeland Security, 153–154
depression metaphor, 63
discourse communities, 12
disenfranchisement, technology of, 154
DNA double helix, black box concept and, 34–35, 43
documents, perpetuity of, 128
double negatives
   handbook entries on, 142–144
   in scientific abstract translation experiment, 186
   social class and, 144
doughnut, conduit metaphor (experimental scenario), 179–182, *179–182*

Ebonics, 130
economy
   communication in agrarian, 47
   knowledge-based, 48
   technical writing as coin of, 90–95, 117
effectiveness, communication
   grammaticality as distinct from, 127–128
   reader's judgment of author's, 141
   Weaver on problem of, 85–86
electronic communication technologies, 76, 82
The Elements of Style (Strunk/White), 136
   recent critiques of, 11
   2000 version of, 16
elite social class, 143–144
employment
   incorrect grammar and, 12, 17
   statistics and demographics, 6, 104–105

encoding, nominalization and, 42–44
engineers, 146–147
　female, 105
English language
　African American Vernacular, 13, 16–17, 126, 130, 143
　exclusion of varieties of, 126–127, 142–144
　legalese and, 4, 23, 190
　Mainstream American (MAE), 11, 129
　nonstandard, 182–189, 191
　oppression tool of, 20
　standardized grammar, 11
English Scientific Register, 102–103, 205n29
　journal articles and, 108, 109
equal opportunity, 133
errors, annoyance from noticing text, 191
ethic of expediency
　human concerns set aside by, 122
　Katz idea of, *31*, 107, 112–115
　Nazi memo and, 112–115
　technological rationalism and, 116
European Scientific Revolution, 91–95
evaluation, in writing style example, 41–42
evolution, of *communication*, 46–49
Excellence Theory, on *communication*, 38
expectations, sentence using, 42
expediency. *See also* ethic of expediency
　clarity and, 107–108
　coding for, 52, 53, 54
　debate over danger of, 206n50
　definition of, 110
　imperative, 50–51
　Plain Style and, 109–110
　purity and clarity for, 106–107
　science journals and, 107–115
　textbooks for technological society and, 156–159

experiments, post-communication metaphor world
　message-passing machine, 178–182, *179–182*
　Molecule X, 169–178, *171*
　nonstandard English and, 182–189
　scientific abstract translation with generative AI tool, 182–189

Fairclough, Norman (CDA), 103, 104
feedback
　loop, *36*
　models to account for, 86
feelings-as-information theory (FIT), 5
Firefox browser, 118
FIT. *See* feelings-as-information theory
Floyd, George, 15–16
Foucault, Michel, 140
"fruitless initially," 147–149, 152

gender
　in employment statistics, 6, 104–105
　Plain Style and, 97, 104–106
　profession conversation anecdote and, 9–10
general communication system, 83
generative AI
　ChatGPT experiment and, 183
　plain language and, 191–192
　scientific abstract translated into nonstandard English with, 182–189
　Standard English fairy tale and, 162–164
good looks, as nominalizing phrase, 43–44
grammar, 10, 15
　job candidates and, 12, 17
　linguists on metaphor and, 199n23
　Lowth and, 136, 143
　Plain English and, 137

Index / 225

grammar *(continued)*
  prepositions and, 16
  standardized English, 11
grammaticality, effectiveness as distinct from, 127–128
Greenfield, Laura, 124, 126, 134, 141–142
  Black Englishes and, 143
  rhetorical thinking and, 155

handbooks, technical writing
  "clarity" and, 138–139, 140
  commodification and, 135–141
  on double negatives, 142–144
  factoid advice approach in, 141–144
  as "plain English," 137
harm, from Communication Metaphor
  from *clarity*, 22–23
  from conduit metaphor, 71–72
  from false beliefs about communication, 71–72
  marginalization of non-SAE communities, 13–14
  metaphoric language and, 24–26
  metaphors power and, 58
  from Standard English fairytale, 124
  suppression of minority voices as, 3–4
  to tribal culture, 18–20
Hawaiian Creole English, 130
The Hid Secrets... See *Arcana Microcosmi*
Hitler, Adolph, 114. *See also* Nazi Germany
Holocaust, 111–112, 206n50
Hooke, Robert, 60–61, 95, 197n33
  piece of cork observed by, 93–94, 94
hyperpragmatism
  human concerns set aside by, 122

software documentation normalized, 118–119
technological rationality and, 115–119, 167–168
hypothetical questions, for teachers, 160–162

IEEE Guide to Writing in the Engineering and Technical Fields, 146–147
Ilsa (tribal college teacher), 19
Industrial Revolution, 115, 116, 120
information
  as nominalization, 121
  technology systems, 191
information theory, 82, 86, 119, 120
  feelings-as-information (FIT), 5
internet, *communication* after invention of, 49
interpretation, problem of, 80, 85
introductory writing course, 2, 29–30

jargon. *See* scientific jargon
job candidates, grammar and, 12, 17

Katz, Steven, 107, 115
  on Aristotle, 112–113, 114, 206n50
Kimmerer, Robin, 183, 187, 210n8
King James Bible, 143
knowledge. *See also* scientific knowledge
  branches excluded by Wikipedia, 7

Lakoff, George, 63
language. *See also* metaphor; windowpane theory (WT) of language
  clear, 78
  controlled, 62, 79–81, 88
  critical language awareness (CLA), 159–162, 209n65

evolution of, 129
exclusion of home, 129, 130
human aspect of using, 157–158
ideology of homogeneous, 127
ideology of standard English, 131–135
legal (legalese), 4, 23, 190
linguistic black boxes, 32–33
linguistic facts of life, 125–130
Locke theory of, 92–93
metaphoric, 24–26
"official," 4
plain, 191–192
roots of scientific, 93
Royal Society and inaccessible, 102–103
Royal Society regulation of scientific, 95–99
standardization, 13
suppression of home, 13–14
technologies of standard, 131, 135
unconscious use of metaphor and, 63
varieties as inferior forms of, 126
written *vs.* spoken, 128–130
language practices, regulation of, 6–7
Large Hadron Collider, at CERN, 46
Latour, Bruno, 34, 35, 198n7, 210n3
legalese (legal language), 4, 23
line, 75, 76, 77, 166
lines of communication, metaphors in, 74–77, *76*
linguistics
  cognitive, 62
  linguistic justice, 129, 134–135
linguists
  black box and, 199n23
  corpora idea of, 46
  critical language awareness (CLA) and, 159–162
  empirical research in vernacular language study by, 130

on grammatical metaphor, 199n23
non-linguists *vs.*, 126, 127–128, 127t
race discussion by, 13
on standard English forms, 11
liposomes
  alternatives for table of, 176–178
  doughnut analogy for, 179–182, *179–182*
  dramatized experiment and components of, *171*
literary metaphor, 64, 87, 177–178
  conduit metaphor in (*line*), 74–77, *76*
  equations created by, 58–61
  line as, 77, 166
  meaning shaped by cognitive and, 58
  STE solving issue of, 80
Locke, John, 91, 92–93, 95, 117
  as colonialist, 106–107
love metaphors, 63
Lowth, Robert, 136, 143–144

MAE. *See* Mainstream American English
Maher, Jennifer H., 119
Mainstream American English (MAE), 11, 129
margin of maneuverability, 190, 211nn12–13
Markel, Mike, 145–146, 149
Marx, Karl, 115–116
masculine norm
  self-censoring in conformity to, 98–99
  stereotypes and, 105
"A Mathematical Theory of Communication" (Shannon), 82
The Mathematical Theory of Communication, 82–83
McClosky, Deirdre, 61

"Measure of Excellence for Technical Documents," 15
memes, nominalization and cultural, 44
message
   conduits, 66, 66–67
   container, 65
message-passing machine, 71, 72, 73, 79
   rhetorical thinking and, 151
   specialist-nonspecialist communication experiment, 178–182, *179–182*
metaphors, 66–77. *See also* analogies; conduit metaphor; expediency
   "clarity" as, 50
   cognitive, 58
   coin, 90–95, 117, 119, 167
   Communication Metaphor and, 24–26, 57
   conceptual, 61–65
   container, 62–65, *65*, 73, 79
   dead, 61
   grammatical, 199n23
   literary, 58–61, 64, 74–77, *76*, 87, 166
   metaphorest fire complex, *31*
   risks and, 25–26
   summary of technical writing, 166–167
   textbooks propagating, 144–148
   unconscious use of language and, 63
Michaud, Michael, 18
Micographia (Hooke), 60–61, 197n33
   cork observation in, 93–94, *94*
microaggression, gender, 10
microscope, Hooke observation of cork under, 93–94, *94*
military-industrial rationality, hyperpragmatism and, 115–119
Molecule X (dramatized experiment scenario)
   replicability warning, 176–177
   saying what is missing, 176
   scenario 1, 169–173, *171*
   scenario 2 (post-Communication Metaphor alternative), 173–178
moon's passage, scientific jargon and, 99–101

National Council of the Teachers of English (NCTE), 12–13
Native Americans
   CEO assignment at tribal college, 18–20
   Locke justification for land appropriation from, 106–107
   violence against, 19–20, 23
naturalized practices, 4
Nazi Germany
   expediency and, 111–115
   technical memo from, 111–112, 114, 115, 158
NCTE. *See* National Council of the Teachers of English
"Next Steps for Families," 153–154
noise, in transmission model, 84, 85, *120*
nominalization
   abstract scientific knowledge enabled by, 99–104
   Bello on cognitive function and, 100–102
   as black boxes, 32–33
   communication as, 8, 32–33, 40–44
   definition of, 57–58
   as encoding complexity into single term, 42–44
   information as, 121
   as style issue, 41–42
non-conduit metaphor statements, 71–72
nonstandard English, 191
   ChatGPT prompts for translating abstract into, 185–189
   scientific abstract translation into, 182–189

objectivity, windowpane theory and, 113

Peters, J. D., 119–121
Philosophic Transactions of the Royal Society of London, 108
phronesis, 157
Plain English, 137, 208n24
Plain Style, 95–99, 117
   consequences for women, 104–106
   expediency and, 109–110
   reimagined, 192
planetary orbit, 43
plant cell
   "cells" and, 25, 60–61, 87, 197n33
   metaphoric language and, 25
post-Communication Metaphor world, 27, 158
   alchemy and, 188, 191–193
   coalitional action framework for developing, 189–191
   experiment 1: Molecule X, 169–178, *171*
   experiment 2: message-passing machine, 178–182, *179–182*
   experiment 3: scientific abstract translation with AI, 182–189
   margin of maneuverability and, 190, 211nn12–13
   Molecule X alternative scenario, 173–178
   Molecule X scenario 1, 169–173, *171*
   putting context in scientific writing, 173–178
   shared and distant contexts, 168–169
   windowpane theory and, 184
pragmatism, progressivism tension with, 159. *See also* hyperpragmatism
praxis, 157

prepositions, 16
present, STE and, 81
professional and technical writing (PTW)
   first day of class in, 22–23
   textbooks, 20, *36*, 36–37
   visibility problem, 30
progressivism, pragmatism tension with, 159
PTW. *See* professional and technical writing
purity and purified
   Hooke and, 95
   justification for clarity and, 106–107
   origin of stylistic, 92–93
   Rickard on, 90–91, 93, 97, 113
   scientific language, 167
   Standard English fairy tale and, 134
put, replace and, 80

questions. *See also* survey responses, on successful writing
   on AI and future of technical writing, 163–164
   hypothetical teaching, 160–162

race and racism
   Black Englishes exclusion, 130
   linguistic scholars and, 13
   rhetorical thinking and textbook example of, 155
   Standard English fairy tale and, 126–127, 130, 134
   writing assumptions based on, 134
rationality, technological, 116, 119–122
readability decline, in scientific journal, 5
readers. *See* audience
red flags, *communication*, 32–33, 42, 54

Reddy, Michael, 67–68, 85
    conduit directness noted by, 74
    lines of communication and, 74
    rhetorical thinking and, 151
    thought experiment of, 69–70, *70*
reductionism, 188
reductionist biology, 61
replace, STE on, 80
research, study of vernacular language in empirical, 130
research articles, origin of, 108–109
Revolutionary War, *line of communication* and, 76
RF. See Rhetorical Framework
Rhetoric (Aristotle), 112–113
rhetoric, Katz on deliberative, 112–113
Rhetorical Framework (RF), 52, 53, 54
rhetorical thinking, 150–156
    audience awareness as, 148–149
    human concerns and, 157–158
    rhetorical persuasion, 140
Rickard, T. A., 90–91, 93, 97, 113
    colonialism and, 106–107
risks, metaphor and, 25–26
Ross, Alexander, 92, 93
Royal Society in England, 92, 95, 137. *See also* Plain Style
    censorship, 97–99, 104–106
    exclusion of women from, 97
    inaccessible language and, 102–103

SAE. See Standard American English
science. *See also* journals; technical writing, scientific and
    Bacon vision of experimental, 95
    censure of audience participation in, 97–99, 104–106
    metaphor in, 25
    origin of research articles in, 108–109
    philosophy of, 188
    readability decline, 5

science, technology, engineering, and math (STEM), 10. *See also* engineers
    diversity and, 6
    group overrepresented in, 4
    majors, 3
Science in Action (Latour), 210n3
science journals
    accepted style in Royal Society, 95–99
    alternative experiment scenario and, 173–174
    black box in academic, *36*, 36–37
    container metaphor and, 145
    English Scientific Register and, 108, 109
    evolution of article genre, 108–109
    expediency and Communication Metaphor and, 107–115
    readability decline, 5
scientific abstract
    Aristotle genre of natural philosophy and, 187–189
    nonstandard English version and ChatGPT prompts, 185–189
    original (in experimental scenario), 185
    translation into nonstandard English, 182–189
scientific jargon
    book on pregnant women example of, 103, 104
    moon's passage example of, 99–101
scientific knowledge
    coin of economy of, 90–95, 117
    nominalization enabling abstract/cumulative body of, 99–104
    shared and distant contexts experiment, 169–170
scientific writing (scientific language)
    abstract, 99–104
    nominalization feature in, 43, 99–104
    nonacademic, 204n22

purified, 167
putting context back in, 173–178
Selber, Stuart A., 145–146, 149
semantic problem, Weaver on, 85–86
severe reason, military-industrial pragmatism and, 115–119
Shakespeare, William, 59, 143
Shannon, Claude, 82–87, *83*, 85, *120*
    technological rationality and, 120–122
Shannon and Weaver model.
    See transmission model, of communication
Sharf, Caleb A., 59–60
shock value, 112
A Short Introduction to English Grammar (Lowth), 136, 143
Simplified Technical English (STE)
    clarity realized through, 79–81
    translation errors, 81
social class, 129, 143–144
social justice action
    communication and, 26–27
software developers, percentage of women, 6, 105
software documentation, hyperpragmatic, 118–119
soufflé, black hole described as, 59–60
Sprat, Thomas, 95
SRTOL. *See* Students' Right to Their Own Language
Standard American English (SAE), 11, 13, 16
    clarity and, 23, 138–140
    home languages and, 129
    home languages excluded by, 129, 130
    as not language, 131–135
    statements about successful writing and, 50
Standard English fairy tale, *31*, 125, 128–130, 182–183

Black Englishes excluded by, 130
definition of, 124
double negatives and, 142–144
factoid advice in handbooks, 141–144
generative AI implications for, 162–164
handbooks commodification and, 135–141
metaphors propagation by textbooks an, 144–148
race and, 126–127
rhetorical thinking and, 148–156
Standard English as not a language, 131–135
Standard Technical English, 87–88
Standard Written English (SWE)
    audience awareness and, 155
    handbooks, 135–141
    home languages and, 129
    Standard English fairy tale and, 131–135
"star death," 60
Starbucks, *150*, 152–153
statements
    about clarity, 138, 139, 140
    codes assigned to, 51–54
    with Communication Metaphor elements, 109–110
    conduit and alternate non-conduit metaphor, 72–73
    container metaphor, 64
    conventional technical writing, 89
    PTW textbooks communication, *36*, 36–37
    revised metaphor, 72–73
states are containers, 63
STE. *See* Simplified Technical English
STEM. *See* science, technology, engineering, and math
Strunk, Oliver, 11, 16, 136, 137

students, technical writing
  CLA assignments for, 162
  common handbook for, 132–133
  expectations of, 124
  Native American, 18–20
  promise to, 156–157
  shock value and, 112
  STEM, 3
  white mainstream college, 19
Students' Right to Their Own
    Language (SRTOL), 12–13
style
  boundary between real world and,
    158
  coding for responses on (S), 52, 53
  coding for style (S), 52
  justification for clarity and purity
    in, 106–107
  masculine norm and censorship,
    98–99
  nominalization and, 41–42
  nominalization as issue of, 41–42
  purity value for, 90–93, 95
  revisions (Plain Style), 95–96
  Royal Society and Plain Style,
    95–99, 104–106, 109
  scientific prose, 98, 204n22
Style (Williams), 137–138
success, as communicator
  conformity required for, 26
  consensus on, 2, 25, 26
  expectation of effortless, 68, 69, 85
  metaphor and, 25, 26
  problem with accepted statements
    about, 50
Sumerian clay tablets, 29, 45
survey responses, on successful writing,
    50
  clarity in, 53, 78, 123
  coding of, 51, 52–54
  collected comments review from,
    123–124

complex answers, 52–54
  on technical and scientific writing,
    2–3
  writing and communication
    agreement between, 1–2, 8, 33–34
SWE. *See* Standard Written English
systems biology, 61
Systole and Diastole, 93

target domain is source domain, 62,
    65
tautology, 39–41, 141
teachers, CLA and technical writing,
    160–162
Technical Communication (Markel/
    Selber), 145–146, 149
technical memo, Nazi van design,
    111–112, 114, 115, 158
technical report conventions, 89
technical writing, scientific and
  as alchemy, 188, 191–193
  coin metaphor for, 90–95, 117
  common statements about, 2–3
  common student handbook for,
    132–133
  communication statements, 36,
    36–37
  conventional statements about, 89
  conversation about profession in,
    9–10
  first writing as, 29
  ghosts eliminated in better writing,
    122
  Katz on deliberative rhetoric and,
    112–113
  nominalization avoidance as
    convention in, 8
  Royal Society and Plain Style,
    95–99, 104–106, 117
  as serving agenda of power groups,
    5–6
  successful, 2, 25, 26

summary of metaphors for, 166–167
windowpane theory and, 78–79
technical writing courses. *See also* handbooks; textbooks
"clarity" and, 138
hypothetical questions for teachers of, 160–162
introductory, 29–30
metaphors in most common, 144–148
Plain Style and, 96–97
student expectations and, 124
technological rationality, 120–122, 167–168
hyperpragmatism and, 115–119
technologies
communication shift with, 47–49
of disenfranchisement, 154
electronic communication, 76, 82
hyperpragmatism, rationality and, 115–119
information, 191
standard languages as, 131, 135
technological imperative, 110
telecommunication, 76
textbooks and, 156–159
transmission model and, 85, 86–87, 112–122, *120*
telegraph, 48, 49, 75–76
terms of use agreement, *150*, 152–153
testing reports, 147–148
textbooks, 78. *See also* handbooks
alternatives to common, 146–147
common student handbook for technical writing, 132–133
container metaphor and, 73
metaphors propagated by, 144–148
1908 example, 90–91
omission of complexity in, 136
pragmatic instructions in, 73
professional and technical writing (PTW), 20, *36*, 36–37

reader-centric, 149–152, *150*
software documentation, 118–119
for technological society, 156–159
tribal college, 19
thought experiment, Reddy, 69–70, *70*
Tohono O'odham Nation, 4
tokenism, 105
transmission model, of communication, 82–83, *83*, *120*, 202n42, 203n46
conduit metaphor (CDM) and, *84*, 84–85
dominance of, 119–122
testing reports and, 148
transportation, *communication* as, 48
treaties, 23
tribal college, CEO assignment at, 18–20

United Kingdom (UK), CLA in, 159–160
USDA Fire Service, 31

violence
against Native Americans, 19–20, 23
Nazi technical memo and, 111–112, 114, 115, 158

waterways, *communication* between, 47
Watson, James, 34–35
Weaver, Warren, 83–87, 120–122
WEV. *See* White English Vernacular
White, E. B., 11, 16, 136, 137
White dominance, 40
White English Vernacular (WEV), 11, 12, 13, 136
clarity and, 23
correctness equated with, 16–17
linguistic justice scholars on, 129
Standard English fairy tale and, 131–135
White Mainstream English (WME), 11, 12, 13, 131

White Mainstream English (WME)
    (continued)
  home languages and, 129, 130
  rhetorical thinking (audience
    awareness) and, 155
  Standard English fairy tale and,
    136
  statements about successful writing
    and, 50
white supremacy, 14
Wikipedia, 7, 13
wildfires, *complex* of, 31, *31*
Williams, Joseph, 137–138
windowpane theory (WT) of language,
    31, 51, 53, *77*, 78, 81–82
  as *clarity* analogy, 21
  clarity statements and, 139, 140
  conduit metaphor and, 77, 140
  grammaticality and, *128*
  objectivity and, 113
  post-Communication Metaphor
    experiment and, 184
  problem with, 79
  statement examples of expediency
    and, 109–110
  STE as fulfilling, 80
WME. *See* White Mainstream English
women
  consequences of Plain Style for,
    104–106
  employment statistics and, 6, 104–
    105
  excerpt from book on pregnant,
    103, 104
  Royal society membership exclusion
    of, 97

World War II, transmission model
    emergence after, 82, 119–122, *120*
writing. *See also* style; technical
    writing, scientific and
    bureaucratic, 45, 190
  first alphabetic letter, 45
  invention of, 29, 198nn3–4
  modern person-person, 45–46
  problem with accepted statements
    about, 50
  regularizing, 136
writing and communication. *See also*
    communication; style; survey
    responses, on successful writing
  clear, 78
  consensus on successful, 2
  fallacy of effortless, 68, 69, 85
  feedback loop, *36*
  incorrect writing as unprofessional,
    15–18, 19
  lines of, 74–77, *76*
  signal *vs.* meaning emphasis in, 121
  specialist-nonspecialist (experiment),
    178–182, *179–182*
  statements in PTW textbooks on,
    36, 36–37
  success without effort concept of, 68
  tautology, 39–41, 141
  technological metaphors for human,
    86–87
  transmission model of, 82–87, *83*,
    202n42, 203n46
Writing Science Right (textbook), 78,
    149–151, *150*, 154–155
  shared goals and, 152
WT. *See* windowpane theory

www.ingramcontent.com/pod-product-compliance
Lightning Source LLC
Chambersburg PA
CBHW021838220426
43663CB00005B/306